W9-CNL-373

symmetry

and the beautiful universe

"Symmetry's significance for the scientific understanding of nature has never before been captured so completely. Leon Lederman and Chris Hill pluck one of the deepest concepts of science from its technical trappings and present it elegantly and clearly, exploring its relevance to everyday life and the depths of the underlying science. *Symmetry* is a grand guide to modern physics and a tribute to the power of beauty to reveal truth."

—Tom Siegfried, Science Editor
Dallas Morning News
Author of *Strange Matters*

"Like many people I have been intrigued with symmetry in the natural and designed world. In this enjoyable and insightful book Leon Lederman and Christopher Hill weave a masterful tapestry as they reveal the depth and breadth of symmetry. Whether a physicist, mathematician, poet, or artist, you will view the world differently after reading *Symmetry and the Beautiful Universe*."

—Rodger W. Bybee, Executive Director
Biological Sciences Curriculum Study
Colorado Springs, Colorado

"From quarks to the cosmos, symmetry shapes the natural world. Hill and Lederman have written a delightful and readable book for anyone curious about how the simple and elegant concept of symmetry has profound implications for the design of the universe."

—Rocky Kolb, Cosmologist
Fermi National Accelerator Laboratory
Author of *Blind Watchers of the Sky*

"Using symmetry as their principal guide, Lederman and Hill take readers on an enlightening tour of modern physics and cosmology. This is a valuable and much-needed perspective."

—Michael Riordan
Author of *The Hunting of the Quark*

symmetry
and the beautiful universe

LEON M. LEDERMAN
nobel laureate

CHRISTOPHER T. HILL

Prometheus Books

59 John Glenn Drive
Amherst, New York 14228-2197

Published 2004 by Prometheus Books

Inquiries should be addressed to
Prometheus Books
59 John Glenn Drive
Amherst, New York 14228–2197
VOICE: 716–691–0133, ext. 207
FAX: 716–564–2711
WWW.PROMETHEUSBOOKS.COM

08 07 06 05 04   5 4 3 2 1

Library of Congress Cataloging-in-Publication Data

Lederman, Leon M.
    Symmetry and the beautiful universe / Leon M. Lederman and Christopher T. Hill.
        p. cm.
    Includes bibliographical references and index.
    ISBN 1–59102–242–8 (hardcover : alk. paper)
    1. Symmetry—Popular works. I. Hill, Christopher T. II. Title.
Q172.5.S9L43 2004
500—dc22

2004008558

Printed in Canada on acid-free paper

*Dedicated by Leon to his teachers at P.S. 93 in the Bronx,
and those in the James Monroe High School*

*Dedicated by Christopher
to his parents, Ruth F. Hill and Gilbert S. Hill*

CONTENTS

ACKNOWLEDGMENTS

For critical comments, insights, shared philosophy, some laughs, and readings of various parts of the manuscript at various stages of development, we thank Shari Bertane, Carol Brandt, Ronald Ford, Stanka Jovanovic, Gilbert Hill, Donald Lorek, Neil Newlon, Laura Nickerson, Irene Pritzker, Bonnie Schnitta, and Susan Tatnall. We also thank our many physics colleagues for their help and insightful comments, especially Andy Beretvas, Bill Bardeen, Roger Dixon, Josh Frieman, Drasko Jovanovic, Chris Quigg, Stephen Parke, and Al Stebbins.

We thank Shea Ferrell, for his fine artwork, and Barbara Grubb, visual collection specialist of the Bryn Mawr College Archives.

We especially thank our patient and dedicated editors, Benjamin Keller, and especially Linda Greenspan Regan, for their heroic efforts in bringing this work to fruition.

We have benefited greatly from numerous people who have visited our Web site at http://www.emmynoether.com and who have commented through their e-mail upon its former contents, which served as a proving ground for the approach taken by this book. And, especially, a hearty thanks to the seven thousand or so students who have matriculated from Fermilab's Saturday Morning Physics Program, to whom some of the content of these lectures was first administered and from whom the inspiration to write this book was drawn. We also wish them the best, since our future depends so critically upon it.

INTRODUCTION
What Is Symmetry?

S ymmetry is ubiquitous. Symmetry has myriad incarnations in the innumerable patterns designed by nature. It is a key element, often the central or defining theme, in art, music, dance, poetry, or architecture. Symmetry permeates all of science, occupying a prominent place in chemistry, biology, physiology, and astronomy. Symmetry pervades the inner world of the structure of matter, the outer world of the cosmos, and the abstract world of mathematics itself. The basic laws of physics, the most fundamental statements we can make about nature, are founded upon symmetry.

We first encounter symmetries in our experiences as children. We see them, we hear them, and we experience situations and events that seem to have certain symmetrical interrelationships. We see the graceful symmetry of a flower's petals, of a radiating seashell, of an egg, of a noble tree's branches and the veins of its leaves, of a snowflake, or of the line of a seashore horizon dividing the sky from the sea. We see the ideal symmetrical disks of the Moon and Sun and their motions in apparently perfectly symmetrical circles through their course in the day- or nighttime sky. We hear the symmetries of a drumbeat or of a simple sequence of tones in a song or in the call of a bird. We witness the symmetry, in time, of an organism's life cycle as well as the symmetry of the seasons, recurring regularly year upon year.

Humans, for thousands of years, have been drawn instinctively to equate symmetry to perfection. Ancient architects incorporated symmetries into designs and constructions. Whether it was an ancient Greek temple, a geometrical tomb of a pharaoh, or a medieval cathedral, each reflected the kind of abode where a "god" would choose to reside. Classical poetry—embodied in such masterworks as the *Iliad*, the *Odyssey*, and the *Aeneid*—invokes symmetrical lyric tempos to celebrate the goddesses or muses of tales and songs. A grand Bach organ fugue, echoing through the rafters of a magnificent cathedral, seems to emanate with a mathematical symmetry, as if reaching down from the vaults of heaven. Symmetry invokes mood, like the sunset on an uninterrupted ocean horizon. The symmetries that we sense and observe in the world around us affirm the notion of the existence of a perfect order and harmony underlying everything in the universe. Through symmetry we sense an apparent logic at work in the universe, external to, yet resonant with, our own minds.

When students are asked to define *symmetry* the answers they give are generally all correct. To the question "What is symmetry?" we hear some of the following responses:

> It's like when the sides of an equilateral triangle are all the same, or when the angles are all the same.

> Things that are in the same proportion to each other.

> Things that look the same when you see them from different points of view.

> Different parts of an object look the same, like the ears or eyes of a face.

These are largely visual impressions of symmetry. Yet we see that they contain a more abstract notion: we see that "sameness" is an ingredient in all of these definitions. In fact, a general definition of the word *symmetry* might be the following:

> **sym'metry** *n.* An expression of equivalence between things.

Symmetry intimately involves the most basic of mathematical concepts:

equivalence. When two things are the same thing, or equivalent, in mathematics, we say that they are equal, and we use the ubiquitous = sign. Thus symmetry is an expression of equality between things. The things can be different objects, or different parts of one object. Or the things can be the appearance of a single object *before* and *after* we do something to it.

A *physical system* is any simple particle, such as an atom, or a complex assemblage of particles, such as a molecule, a rock, a person, a planet, the whole universe, that moves or behaves according to the laws of physics. Essentially everything becomes a physical system, when viewed through the prism of physics. A physical system is said to *possess a symmetry* if one can make a *change* in the system such that, *after the change, the system is exactly the same as it was before.* We call the change we are making to the system a *symmetry operation* or a *symmetry transformation.* If a system stays the same when we do a transformation to it, we say that the system is *invariant under the transformation.*

So, a scientist's definition of symmetry would be something like this: *symmetry is an invariance of an object or system to a transformation.* The *invariance* is the sameness or constancy of the system in form, appearance, composition, arrangement, and so on, and a *transformation* is the abstract action we apply to the system that takes it from one state into another, equivalent, one. There are often numerous transformations we can apply on a given system that take it into an equivalent state.

A simple example of a geometrical symmetry is provided by a Chinese flower vase before the decorations are glazed on it. If the vase is sitting on a table, and we rotate it through any angle (perhaps 37.742 . . . degrees), nothing about its appearance or physical makeup changes—the "before" and "after" photographs of the vase are identical. The vase is invariant under any rotation about an imaginary line in space coursing through the center of the vase, which we call the *symmetry axis.* This simply affirms that our mathematical definition of symmetry coincides with our perceptual experience, and the emotional one as well, that symmetry enhances the beauty endowed in the vase in its form and shape.

SYMMETRY IN MUSIC

Let's think of symmetry in regard to something that is familiar but not necessarily visual. As we have noted, symmetry is ubiquitous, and it permeates art, including one of the grandest of art forms, music.

Western music in the era of Johann Sebastian Bach was progressing beyond the earlier baroque styles, which were comparatively simpler in their outlines, inherited from the Renaissance. Music was emerging into a new era, in which there was more feeling, emotional content, and moods—called *Affekt*. Moreover, the form, structure, or architecture of music was undergoing an evolutionary change.

Bach, as a boy of fifteen in the year 1700, won a scholarship to study at the Michaelisschule in the city of Lüneburg, about thirty miles from the northern city of Hamburg, in what is today Germany.[1] He was granted free tuition, room and board, and an allowance for his services as a church choirboy, including performances in Sunday services, weddings, funerals, and various festive occasions. He sang as a soprano, but the scholarship expired after a few years, together with his career as a student, when his voice changed.

The city of Lüneburg provided a stimulating and diverse musical and intellectual life for a young music student. Here Bach encountered for the first time a new "symmetrical" style of composition, which was seen earlier in the structure of music of the French composers of the era, composers such as François Couperin. Music, through these composers, was becoming more humanized, intimate, subtle, and ambiguous, increasingly representing everyday human activities such as the courtship ritual—dance—in its structure and form. And, like dance, music was acquiring more symmetry.[2]

A simple, regular drumbeat is a repetitive rhythm, a symmetry in time. The rhythm of a drumbeat measures out time in equal intervals. By our definition of symmetry, the *equality of the interval* between drumbeats is the *invariance*, with the *passage of time* being the thing that is changed, the *operation* or *transformation*. The physiological symmetry of the heartbeat is another example. Arrhythmia, therefore, is asymmetry. The drumbeat represents the beat of the heart, the rhythm of life. Music evolved from the rhythm of a drum.

An early composition would typically have a theme, which we shall call X, that played on and on, repeating itself, in a given key. Consider the well-known and popular Pachelbel Canon in D. Johann Pachelbel was born just a quarter of a century before Bach and participated in the eighteenth-century expansion of the musical vocabulary. Pachelbel's Canon in D illustrates the symmetry of earlier baroque music. The canon takes the form of a theme, consisting of the chord progression D–A–Bm–F♯m –G–D–G–A, in a continuous, almost clocklike tempo, repeating itself over and over, with clever variations and tantalizing embellishments, as different voices exit, enter, and harmonize.

There is, of course, nothing wrong with this form—modern composers still use it to evoke the mood of ephemeral motion, such as Ravel's *Bolero*, of the twentieth century, a canon evoking a feeling of a steady onward march of events, setting the stage for a climactic finale. But, during the time of Bach, music was beginning to evolve a more complex symmetry pattern, representing the first *compound musical form*.[3] Such compositions contained structures, called *movements*, which were imitative of dances, themselves imitative of actions seen in nature. Such movements were called *allemandes, courants, sarabandes, gigues*, and *fugues*, named after popular dances of the era. These movements in a composition followed a rigorous set of rules, defining their symmetrical pattern.

Now the first movement, X, would have been the main theme, written in the defining, or tonic, key of the composition, and would change, or modulate, into the dominant key (for example, the tonic key of C major modulates into G). This was followed by the second movement, Y, which would be the continuation of the same theme in the dominant key, modulating back into the tonic key (the key of G major in our example, ending back in C). This XY, or binary form, structure was expanded into other patterns throughout various compositions, such as the XXYY, which is called the repeated binary form. Later forms, such as found, for example, in the piano sonatas of Beethoven—the so-called Viennese allegro-sonata style—are generalizations of this basic symmetry pattern, where Y may be X restated in a related key other than the dominant key, perhaps the relative minor key (for example, if X is stated in the tonic key of C major, then Y would be a restatement of X in A minor), Y usually containing thematic variations on X.

Bach absorbed these novel concepts, but he infused music with much more symmetry than these mere outlining patterns. In many of the movements of Bach's compositions there are symmetrical subdivisions into what are called phrases and semi-phrases, with similar patterns that reflect and imitate the outlining structural symmetry of the piece. One further finds in Bach's compositions a hallmark feature, called the back and forth technique (mentioned earlier), where similar measures in sections X and Y employ the same themes but with a reversed sequence of tones. The individual musical phrases themselves form symmetrical subcomponents of the larger synthesis. The overall composition becomes a hierarchy of these symmetrical components, with a diversity of things on many different scales of time and space.

Listeners often cannot grasp a Bach composition upon the first hearing; it requires patience, often several hearings, before we begin to comprehend the inner world of these majestic compositions, in which this complex hierarchical structure takes wings and soars. As we begin to comprehend, we feel as though we are experiencing a new and complex universe, with unfolding patterns upon patterns, a universe defined by underlying principles of logic and symmetry. The music transcends the instruments upon which it is played. Bach sounds as "right" on a kazoo or an electronic synthesizer as on a harpsichord or a massive pipe organ. It is ultimately not the particular instruments that define the structure of music, but rather the deep internal symmetry structures themselves and the overall *Affekt* they produce.

EARTH IS ROUND

Symmetry gives wings to our creativity. It provides organizing principles for our artistic impulses and our thinking, and it is a source of hypotheses that we can make to understand the physical world. Let's turn to another magnificent example of this: the discovery that Earth is round. This did not await the second millennium, with the voyages of Columbus or Magellan and the first physical circumnavigation of the globe. Magellan performed the "confirming experiment" to prove the theory (though he himself did not survive the trip, a consequence of a failed attempt to convert the Philippines to Christianity[4]). Rather, Earth was already known by ancient Greek mathematicians to be a sphere, like the Moon and the Sun—and they had measured its diameter.

The Greeks had noticed that on occasion, Earth blocks the sunlight from hitting the Moon, causing what is called a lunar eclipse. By observing the shadow of Earth cast upon the Moon during a lunar eclipse, they could see that Earth was also a round body, a sphere, just like the Moon and the Sun.

Eratosthenes, a Greek scholar and the chief of the famous ancient library of Alexandria, Egypt, around 240 BCE, knew that in a town far to the south, Syene, there was a deep water well. On the summer solstice, the longest day of the year—June 21—the full image of the sun could be seen reflecting, for a brief moment, in the water of the deep well in Syene at precisely noon. Therefore, the Sun at noon must be passing exactly overhead in Syene. He noticed, however, that on this same day, the Sun did not pass directly overhead in his hometown of Alexandria, which was 800 km (500 mi) due north of Syene. Instead, it missed the zenith, the point directly overhead in the sky, by about seven degrees. Erastosthenes concluded that the zenith direction was different by seven degrees in Alexandria from that in Syene. Using some elementary geometry, he could then determine the diameter of Earth and found it to be 12,800 km (8,000 mi).[5]

Earth's true diameter, as we know it today, depends slightly upon where you measure it, since Earth is *oblate*, that is, wider through the equator than through the poles, and it also has mountains, tides, and so on, that require us to quote only an "average value." The average diameter of Earth through the equator is about 12,760 km (7,929 mi), and through the polar axis, about 12,720 km (7,904 mi). This means that Erastosthenes derived the correct result for Earth's diameter to an astounding precision of better than 1 percent, assuming Earth was a sphere. This was a remarkable scientific achievement for that time.

Yet, in fact, Earth, as noted above, is *not* a perfectly symmetrical sphere as envisioned in the ideal abstract geometrical limit. The symmetry of a sphere is only an approximation to the planet's shape, which is determined by the dynamic processes of matter accreting to form a large, solid body under the influence of gravity. It would, for example, be wrong to conclude that it was a divine hand that created Earth as a perfect sphere and to thus associate it with a religious "perfect sphere" belief system.

Symmetry can be a powerful tool, even when it is only an approximation to reality. But our human species has often made mistakes, assuming some things have or are perfect symmetries when the symmetries are actually only illusory or accidental consequences of something else. This was the mistake of Ptolemy's theory of an Earth-centered solar system, which held sway hand-in-hand with religion for a millennium and a half. The symmetries of the perfect circle and of the sphere were assumed to be divine, something designed by God, which meant they simply *had* to be there, expressed directly in the orbital motions of the planets, the Sun, the Moon, and the stars about the supposedly fixed Earth at the center.

Indeed, there are symmetries in the motions of the planets—but the true symmetries are hidden and more profound than anyone at that time could have imagined. Johannes Kepler had the wisdom and the perseverance to find the exact principles describing planetary motion around the Sun. These principles seemed disappointingly imperfect, far adrift from the preferred mandate of spherical symmetry and geometry. Nevertheless, they set the stage for the greatest intellectual run in human history, from Galileo to Newton to Einstein, and to the ultimate unveiling of nature's deepest and most profound symmetries.

SYMMETRY IN MATHEMATICS AND PHYSICS

Mathematicians have evolved a systematic way of thinking about symmetries that is fairly easy to grasp at the outset and a lot of fun to play with. This almost magical subject is known as *group theory*. It began with nineteenth-century French mathematician Evariste Galois, who in his short, tragic life laid the foundation for this way of thinking.

Galois was a political radical, and he became tempestuously involved with a beautiful woman who was already engaged to a man named Pescheux d'Herbinville. D'Herbinville was known to be a fine marksman, discovered the liaison, and challenged Galois to a pistol duel.[6] The night before the duel, Galois, knowing the reputation of his opponent, frantically scribbled down notes, summarizing his mathematical analyses of higher-order algebraic equations (typically quintic, or those of the fifth order), and a method that determined their solvability. At the heart of this analysis is the algebraic structure of group theory. On the morning of May 30, 1832, with one shot, the twenty-one-year-old Galois was felled and left to die on the "field of honor." It has been alleged that, in fact, the duel was a setup to assassinate the politically extreme Galois. Fortunately, however, some fourteen years later, Galois' notes came into the hands of the eminent French mathematician Joseph Liouville, who recognized their brilliance and significance and communicated them to the world.

Group theory is the mathematical language of symmetry, and it is so important that it seems to play a fundamental role in the very structure of nature.[7] It governs the forces we see and is believed to be the organizing principle underlying all of the dynamics of elementary particles. Indeed, in modern physics the concept of symmetry serves as perhaps the most crucial concept of all. Symmetry principles are now known to dictate the basic laws of physics, to control the structure and dynamics of matter, and to define the fundamental forces in nature. Nature, at its most fundamental level, is defined by symmetry. The picture we have now, which was constructed gradually, mostly throughout the twentieth century, is still incomplete. There are, however, enough pieces of the jigsaw puzzle in hand to know that symmetry is fundamental to all of it. The abstract concept of symmetry and its relationship to the physical world is enduring and here to stay.

In the midst of the fomenting of the new twentieth-century physics was the ascetic and somewhat tragic life of the greatest female mathematician who ever lived, Emmy Noether. She practiced in the center of the intellectual universe of her age, the University of Göttingen, Germany. There she worked with the greatest mathematician of the age, David Hilbert, and through her work greatly influenced Albert Einstein. She pioneered an academic role at odds with what was normally assigned to women and ultimately bore witness to the collapse of European civilization. She broke through an almost impenetrable glass ceiling to become a university lecturer, only to be subsequently dismissed from the university for being a Jew. She poignantly bid farewell to friends and family, whom she would never see again, and spent the remaining few years of her life at Bryn Mawr University in Pennsylvania.

At Göttingen, Noether achieved fame for her research into the fundamental structure of mathematics. However, she stepped briefly into the realm of theoretical physics to prove a remarkable mathematical theorem about nature. Noether's theorem is a profound statement, perhaps running as deeply into the fabric of our psyche as the famous theorem of Pythagoras. Noether's theorem directly connects symmetry to physics, and vice versa. It frames our modern concepts about nature and rules modern scientific methodology. It tells us directly how symmetries govern the physical processes that define our world. For scientists, it is the guiding light to unraveling nature's mysteries, as they delve into the innermost fabric of matter, exploring the most minuscule distances of space and shortest instants of time.

To this task scientists apply the most powerful microscopes humans have built. These are the great particle accelerators, such as the Tevatron at Fermilab in Batavia, Illinois, and the Large Hadron Collider under construction in Geneva, Switzerland. The Tevatron accelerates protons and antiprotons in opposite directions in a great circle, to energies of one trillion electron volts, as if one had a one-trillion-volt battery hooked up to a vacuum tube. These particles then collide head-on. The quarks and antiquarks, inside the protons and antiprotons, themselves collide. By reconstructing the debris from a collision, physicists get a kind of "photograph" of the structure of matter at the shortest-distance scales ever seen, distances as small in comparison to a basketball as a basketball compared to the orbit of Pluto! These collision events reveal the basic constituents of matter and the fundamental laws of physics that govern their behavior. We find that this behavior is governed by symmetry.

By studying physics at minuscule distance scales we can see that the forces of nature begin to merge together and share a common property, a phenomenon that is unseen at a lower energy, or "magnification." Today we have come to understand that this merging together, or unification, of the basic forces into one entity is a consequence of a single, elegant underlying symmetry principle. This principle, called *gauge invariance*, is subtle. Armed with this principle, scientists can now contemplate the universe at the earliest instants of time. Out of the crucible of quarks, leptons, and the fundamental gauge forces has come modern cosmology.

The discovery of the unifying symmetry principle of gauge invariance has allowed us to leap, theoretically, to distance scales one thousand trillion times smaller than can be seen with our most powerful particle accelerators. This has also allowed us to conceive of what the universe itself was like in the first one-millionth of one-billionth of one-billionth of one-billionth of a second! At such short distances, about $1/1,000,000,000,000,000,000,000,000,000,000,000$ of an inch (that's a one divided by a one followed by thirty-three zeros, or, written in the more convenient scientific notation, 10^{-33}), quantum gravity becomes active, space and time break down, and our normal notions of reality are obliterated. There we must try to use the symmetry principles (and related mathematical ideas, such as *topology*, the study of the possible shapes and forms of surfaces) to try to conceive theoretically where the complete unification of all forces will ultimately lead.

Such research has led to remarkable new ideas, to *superstring theory*, with its mysterious overarching mathematical system called *M-theory*, which no one yet fully understands (let alone what the *M* stands for). Nevertheless, this is perhaps the most symmetry-filled logical system ever conceived by the human brain, and it is our best hypothesis for getting to the so-called theory of everything in the physical universe. Or else, like Ptolemy's solar system, perhaps this effort has yet to identify the true and hidden symmetries of nature, to be unveiled by the next Kepler.

To understand where our ideas about symmetry in science begin, let us begin at the beginning. Let us roll back the clock and consider a time when the universe was still very young but appeared to be a burned-out failure—a dud: nothing of substance existed in it, nor did it appear likely that anything other than some pointless clouds of hydrogen gas ever would. How did we get from there to here?

Let's examine the history of our universe and the history of our particular planet, as modern science now understands it. We do this through a prism—the prism of early Greek mythology, glimpsing humans struggling to understand the very concept of "origins." We start with the relatively late early universe, somewhere around its ten millionth anniversary. Side by side we'll consider the account of the origin of planet Earth and of *us*, as seen through mythology and science.

Mythological stories, created by humans in lieu of the insights of scientific observation, assign human traits to the forces of nature. The scientific history of the universe, on the other hand, has been divined from countless experiments, observations, and measurements, using telescopes and microscopes (particle accelerators), ultimately synthesized into mathematics. Here we'll see a blending of the power of physics with poetry and tradition, contrasting and coalescing, ultimately synthesizing our modern understanding and methodology.

Our intent is to show that this photograph of knowledge—here sharp and well focused, there still fuzzy, and way over there still shrouded in total mystery—is nonetheless governed by a universal and steadfast set of *laws of physics*. These laws are not yet completely understood, but they endure, govern, and control the awesome history of the universe itself. There is solid scientific evidence that the laws of physics are unchanging, derived in part from the geological record of the early Earth. They are the same laws of physics today that supervised the very early universe. This unchanging, or invariant, set of laws is constructed from profound intrinsic symmetries, and they act to express the awesome beauty of nature.

A Tribute to Emmy Noether

Emmy Noether's work interweaves our understanding of nature—through physics and mathematics—with the beauty and harmony that surrounds us in all forms, in nature, music, and art. Emmy Noether made one of the most significant contributions to human knowledge through her remarkable theorem. The theorem cleanly and clearly unites symmetry with the complex dynamics of physics and forms a basis for human thought to make forays into the inner world of matter at the most extreme energies and distances. One might argue that Noether's theorem is as important to understanding the dynamical laws of nature as is the Pythagorean theorem to understanding geometry.

In fact, Noether's theorem provides a natural centerpiece for any discussion that unifies physics and mathematics, such as in the teaching of these challenging subjects in a way that enlivens them both. Her insights offer an approach that makes not only a single lecture but the entire lower-level curriculum of physics, math, and other sciences breathe new life. It explores new concepts in mathematics, symmetry groups, and pulls math back into the science class, where it sits comfortably.

The brilliant contributions of Emmy Noether also have an important sociological benefit—here was a genius, and probably the greatest woman mathematician in history. Very few students, indeed few people, have ever heard of her, yet here is a consummate role model, for women, for everyone, whether they are interested in becoming scientists or not.

Yet, most young girls enrolling in a high school or college physics course must feel, on the first day of class, that they have inadvertently stumbled into the men's locker room. Galileo, Newton, Einstein, Heisenberg, Schrödinger, Fermi—the panoply of heroes of physics is not gender-balanced like the panoply of gods on Mount Olympus, or the characters in Shakespeare's plays, or Italian operas. It is little wonder that few girls pursue the subject. Physics, however, should not be a men's club—a "bathhouse," as eminent mathematician David Hilbert described the attitudes of fellow faculty members while fighting for Noether's well-deserved yet reluctantly awarded promotion to professor; there is no difference in ability and insight that dedicated people of any gender can bring to bear in any intellectual pursuit.

Although today we count an increasing number of young women becoming physicists, the numbers are still incredibly small. We sadly note that despite the awesome role model of Emmy Noether, as well as Marie Curie, Catherine Herschel, Sophie Germain, and many other great women of science and mathematics throughout history, in the year 2005, women will still be greatly underrepresented in the physical sciences, especially in mathematics and physics. Deep cultural biases evidently persist into the twenty-first century. The scientific community can no longer tolerate or afford such a talented group to be so underrepresented.

The perspective offered by symmetry provides a vehicle to invigorate the centuries-old Galilean-Newtonian physics. It provides an orientation and road map to modern thinking about nature and the avant garde, Einstein's relativity, and the unification of all forces under gauge symmetry. It defines the road to superstrings. Therefore we don't hesitate to write a popular book of physics from this perspective. And this goes hand in hand with our desire to see a better physics course taught in high school or lower-level college.

The world we live in today is overwhelmingly complex, and we all face challenges that are more difficult and urgent than ever. The tools we could use to solve the world's problems require basic research and advanced technologies. The issues surrounding the science that underlies this often go well beyond the grasp of the voting public. Therefore we must strive to counter the declining awareness, participation, and understanding of the technological fields of science, engineering, and mathematics. It is imperative that we try to give the nonscientific members of society, who, through democratic processes, make the final decisions, a better understanding of the key issues. In fact, our future depends upon it.

Above all, perhaps, the life of Emmy Noether and her difficulties as a woman in science provide a timely lesson in the need for tolerance and diversity in our society and the pursuit of truths.

chapter 1

CHILDREN OF THE TITANS

.

The Titans are slain by Zeus' thunderbolt;
but out of their ashes man is born.
—Arthur Koestler, *The Sleepwalkers*

THE EVOLUTION OF THE UNIVERSE AND ITS METAPHORS

Ten million years after the big bang, a mist of particles filled the universe. A thin fog permeated space, containing only the lightest of atomic elements, mostly hydrogen and some helium gas. There were also a few species of elementary particles, remnants of the ferocious instant of creation, roaming freely through space. It was dark and becoming cold, lit by only a faint infrared glow—the relic radiation of the big bang, like the glow of the cooling ashes of a dead fire.[1] By its ten millionth anniversary, the universe appeared to be dying.

The universe contained no materials out of which to make solid objects. It would appear that there could never be *things*, such as seashells, trees, icebergs, statues of David, freeways, guitar strings, feathers, brains, stone implements, or paper on which to compose an original Bach cantata. Indeed, there could be no rocks, or sand, or water, or a breathable planetary atmosphere, much less a planet. No solids could possibly form out of the diffuse gases or the fleeting elementary particles, adrift and marooned within the immensity of space. At ten million years, a very short time for a planet, or even for an entire species of life on Earth, the universe was thus formless, cold, dark, and, apparently, just fading away.

For reasons that are not yet fully understood today, perhaps having to do with one of the mysterious, perhaps as yet unknown, species of elementary particles present in the primordial fog, *something* did happen. It may have been little more than the spontaneous formation of small clumps of particles, stirred by quantum motion, forming tiny primordial seeds of structure, like the seeds of dust that cause water vapor to coalesce into drops of rain over the plains of Kansas. But it was enough to set gravity to work. By the uncheckable and invincible force of gravity, parts of the mist began to collapse into gigantic clouds. The great hydrogen clouds began to swirl and roil, like massive thunderheads. The gravity-fed collapse became more intense. Within a few hundred million years, a complete transformation of the formless mist had occurred. Large, primitive, blob-shaped galaxies, each containing billions of faint and youthful stars, began to shine. The universe began to bloom.

These first stars were the parents and grandparents of everything to come. Some were barely more than enormous soft balls of hot hydrogen gas, hardly able to glow. Others became superstars, enormous brilliant spheres, hundreds of times as massive as the Sun, shining radiantly blue as they savagely devoured their primordial fuel of hydrogen and helium. Deep within the cores of these titanic stars, heavier atoms formed, built up from the hydrogen and helium fuel through the process of nuclear fusion.

The extreme pressures and temperatures found deep within the interiors of stars foster the process of nuclear fusion. The joining together, or fusing, of atomic nuclei, makes heavier atomic nuclei. A pair of helium nuclei squeezed together make a beryllium nucleus; add another helium nucleus, and a carbon nucleus is created; a carbon nucleus plus a helium nucleus yields an oxygen nucleus; and so forth. This process produces the energy that fuels the star, causing it to shine brilliantly, emanating its intense radiation of light into the dark void of the universe.

The sequence of fusion proceeds, manufacturing ever-heavier atoms within the nuclear furnace of the stellar interiors, until it reaches the element iron. Iron is the most stable atomic nucleus and, together with those of the heavier elements, will not yield more energy by fusing with other atomic nuclei. Iron marks the end of the available fuel of a star and the ultimate end of the life of a star. The smaller stars, upon exhausting their fusion fuels, simply cease to shine, shrinking down into cold, dead worlds, sleeping eternally and invisibly within the galaxy. The superstars, however, experience a far more dramatic and violent fate.

THE TITANS

All civilizations seek to understand what awesome forces, rules, or laws drove the sequence of events from which the physical world materialized. By whom or by what canon is an entire universe created? In what language must the story be told? Can all of the questions ever be answered?

This history of the evolution of the universe, from the initial explosion to the creation of galaxies, containing billions of clustered stars shining in the darkness, was deduced by humans—who themselves are the product of a quite different scale of evolution, taking place on a unique planet, yet orbiting a typical star, part of an ordinary galaxy. This account is the scientific one. Yet it is an illuminating lesson in the development of human thought to reflect upon the *evolution of the idea of creation*. There are conceptual seeds of our modern cosmology to be found even within the pagan myths of the ancients, such as those of the early Babylonians, Egyptians, and Greeks. From them we can fathom how the early rational mind grappled with the profound logical puzzles that the universe poses.

We have come, in our time, to systematize our understanding of the rules of nature. We say that these rules are the *laws of physics*. The language of the laws of nature is mathematics. We acknowledge that our understanding of the laws is still incomplete, yet we know how to proceed to enlarge our understanding by means of the "scientific method"—a logical process of observation and reason that distills the empirically true statements we can make about nature. The "logical process," we note, is often clouded with uncertainty, hounded by confusion, tripped by errors, delayed by bureaucracy, and stymied by egos, but in the long run, the logical process wins out. Scientists thus strive to determine the steadfast laws of nature. While we believe today that the established laws of physics permeate the universe so completely that they were the same laws at the instant of creation as they are today, this is nonetheless a scientific hypothesis, for which scientists constantly seek observational confirmation.

Similarly, the ancients sought a system of steadfast rules that justified their view of creation. The ancients' conception of the forces and laws governing creation were also based upon the empirical observation of the world around them. Their rules, however, were the "laws" of human nature and the "rules" of human emotion, which included the foibles of human behavior. These behavioral traits were projected onto their gods, the prime movers of the universe. Rather than the abstract language of mathematics, their language was poetry.

The Titans of the ancient Greek pagan creation mythology are, in a bizarre way, the metaphorical analogues of the first large stars that formed in our universe—stars that ultimately became supernovas. The Titans were the mysterious first generation of the gods, called the "elder gods," the parents and grandparents of the later gods of Mount Olympus. There are many gods in this tale that personify a broad range of human traits. Hence, the story is rife with bawdiness, love, promiscuity, incest, pillage, resentment, envy, jealousy, violence, and all the other stuff of a nine-teenth-century opera. In this story we also find a unique logic similar to that of the modern scientific account of creation.

According to Greek mythology, Chaos existed before the Titans. In the era of Homer—the eighth century BCE—the poet Hesiod wrote in his *Theogony* that the goddess Gaia (Earth) spontaneously emerged from Chaos and gave birth to Ouranos (Heaven—*Uranus* in Latin). We inherit Gaia as the prehistoric "mother-earth" goddess worshiped by ancient Western tribal cultures before the rise of the Hellenistic civilization:

> Verily at the first Chaos came to be, but next wide-bosomed Gaia [Earth], the ever-sure foundations of all the deathless ones who hold the peaks of snowy Olympus, and dim Tartarus [Hell] in the depth of the wide-pathed Gaia, and Eros [Love], fairest among the deathless gods, who unnerves the limbs and overcomes the mind and wise counsels of all gods and all men within them. From Chaos came forth Erebus [Gloom] and black Night; but of Night were born Aether and Day, whom she conceived and bore from union in love with Erebus. And Gaia first bore starry Ouranos [Heaven], equal to herself, to cover her on every side, and to be an ever-sure abiding-place for the blessed gods. And she brought forth long hills, graceful haunts of the goddess-Nymphs who dwell amongst the glens of the hills. She bore also the

fruitless deep with his raging swell, Pontus, without sweet union of love. But afterwards she lay with Ouranos and bore deep-swirling Oceanus, Coeus and Crius and Hyperion and Iapetus, Theia and Rhea, Themis and Mnemosyne and gold-crowned Phoebe and lovely Tethys. After them was born Cronos the wily, youngest and most terrible of her children, and he hated his lusty sire.

Thus Gaia incestuously mated, with her first son Ouranos, the offspring that he proudly named the Titans, beings of enormous size and incredible strength. The most famous of the mythological Titans include *Cronus*, the Roman Saturn, the father of Zeus; *Oceanus* (the sea); *Mnemosyne* (memory); *Tehemis* (justice); and *Iapetus*, whose son Atlas carried the world on his shoulders. *Prometheus* was a Titan who stole fire from the gods to save the race of humans and inspired human endeavor to comprehend the universe. *Tartarus* personified the underworld in Hesiod's poem, a dark, gloomy, and forbidding place, the original Hell, surrounded by a great iron fence. It was the ultimate prison for all who arrived, its gates guarded by the most hideous creatures in the universe. Tartarus was considered to be "below all things," though his gates could be reached by jumping into a volcano, falling for nine days. The Titans were the parents of the gods who ultimately ruled from Mount Olympus. All others in the Greek mythology are descendents of the Titans.[2]

We can draw some tantalizing parallels between our scientific account and the ancient myth, though of course Hesiod could not have foreseen them. For example, Gaia's dark sibling, Tartarus, can represent the gigantic black holes that are believed today to lie at the centers of many of the galaxies. These were formed as the primordial cloud of gas crushed in upon itself, yielding the first structures in the universe. The trip down the volcano to Tartarus could be likened to a poetic description of the one-way trip of an unfortunate spacefarer as he falls through the *event horizon*, the boundary of a gigantic black hole, never to return to his own universe and home again. Imprisonment is truly everlasting in a black hole once one crosses the event horizon, much stronger than any iron gate guarded by even the most ferocious of hideous monsters. There prevails a rearrangement of space and time, from which not even light itself can reemerge.

The era of Hesiod, like the early European Renaissance, was a period of the flowering of literature, but in the so-called heroic age of Greek civilization. And, as occurred after the Renaissance, there ensued a more analytical or rational era, an "enlightenment," yielding the development of mathematics. In ancient Greece this occurred with the rise of the school of the great sixth-century BCE mathematician Pythagoras. This was a time and place wholly unique in human history, when the refined human mind first realized that mathematics describes the physical world.

With the new tool of geometry in hand, Pythagorean philosophers tackled structural questions about the universe. They asked, given the logical order of mathematics, how is the universe put together, such that it embodies this logic? What is its shape? How do its components move? What is the (atomic?) composition of all matter? Is Earth at the center of the universe, and if so, how do we reconcile the observed motions of the planets in the sky? The Greeks thus perfected geometry and logic and developed detailed scientific theories of most natural phenomena, including the tides, weather, the origin and evolution of the species, medicine, matter, and the cosmos.

This remarkable enlightenment may have silently culminated in about 310 BCE in a scientific and theoretical masterpiece by a brilliant philosopher, Aristarchus. Building upon a Sun-centered theory of the solar system proposed earlier by his predecessor, Herakleides, Aristarchus extended the theory to describe correctly the true configuration of the orbits of Earth and the other planets encircling the Sun, as well as that of the Moon, encircling the Earth. This work was lost but is nonetheless known to us, having been described by the Greek scientist Archimedes and the Roman-era philosopher Plutarch. This may represent, symbolically, the high-water mark and end of the golden age of Greek scientific philosophy, an era that has been described as a mere step away from Copernicus, Kepler, and Galileo.[3]

The Sun-centered theory was seen by some as freakish and was never accepted by later Greek philosophers. (This essential key to unlocking the laws of physics would have to await its rediscovery by Copernicus and Kepler almost two millennia later.) The nature of philosophy itself changed, the reverence to mathematical and scientific rationalism declined, and society underwent upheavals, leading to the age of Plato and Aristotle. These philosophers got the whole picture of the structure of the universe wrong, leading ultimately to the widespread acceptance of misconceptions about physics and natural phenomena. This ultimately became canonized in the doctrines of the authoritarian Catholic Church.

Despite its remarkable achievements, during the age of the Pythagoreans, the detailed understanding of the cosmic *origin* of the universe progressed little beyond poetic allegories such as that of Hesiod. Of course, meaningful scientific observations of deep space were unavailable in that age. Remarkably, however, the pagan creation myth does resolve for us a logical question about creation—and it gets the answer right! It adopted the correct idea of a *singular* tumultuous creation event, that is, the universe sprang from chaos, an ill-defined nothingness, similar in its broadest strokes to our modern theory of the big bang.

How can there be such a striking parallel between an ancient myth and a modern scientific theory of creation? In actuality, there aren't many options. Any creation story is essentially the solution to a logical puzzle. Either the universe has always been here, in which case the question of creation becomes a moot point, or else it was created at some particular, singular instant of time. A third possibility, perhaps more Zen-like in its viewpoint, is that reality is an illusion and the universe as we know it was never created in any meaningful way—that is, perhaps the question itself makes no sense. The Greek creation myth solves this puzzle by insisting upon a singular event of creation, confronting head-on the task of "explaining" the unique event. The ancients' explanation is also an attempt to understand the violent and detailed processes of creation through the underlying "laws of nature," albeit meaning the laws of human emotion, the tempestuousness of the gods, and their gods' wild behavior. This engaging story depicts our most human qualities, both good and evil. The subsequent devolving logical sequence of things led ultimately to the planet Earth that we inhabit today.

Only within the past forty years or so has modern science come to the consensus that there was an initial instant of creation, the so-called big bang. Whereas Hesiod's myth began at the top of the mountain and carried the poetic tablet down from on high, science, with the scientific method, has had to climb the mountain, arduously, through a long and tortured history of painstaking discovery, analysis, refutation, and ultimate success. Getting there was not easy. It involved detailed understanding of scores of fundamental processes and observations. Discoveries such as the observation of the three-degree Kelvin background radiation (the leftover electromagnetic radiation from the big bang that persists today) are among the immediate scientific discoveries that confirm the theory, and many recent discoveries have significantly enhanced even further our detailed picture of what happened. But our picture of the creation of the universe rests upon *all* of the discoveries of the science of physics. Indeed, we have learned perhaps more about the cosmos by looking through the world's most powerful microscopes—the particle accelerators—as by looking through telescopes. There is no doubt that there was a singular instant of creation, the big bang, occurring approximately fourteen billion years ago. Our planet Earth, in actuality, developed rather late in the true sequence.

According to our modern scientific view, the universe emerged from a "chaos" of matter, a plasma of the elementary constituents of matter—quarks, leptons, gauge bosons, and many as yet undiscovered particles—furiously swarming about at extreme temperatures and pressures in an embryonic warped and twisted space and time. Space itself exploded, driven by the raw energy of the constituents of the universe, as later explained by the geometrical laws of Einstein's general theory of relativity. As the universe and its constituent plasma expanded, it cooled and condensed, ultimately transforming itself into normal matter, forming a uniform gas of hydrogen, some helium, and relic particles of electromagnetic radiation, neutrinos, and perhaps some unknowns. Primordial quantum fluctuations in the density of these relic particles may have been transmitted, through gravity, to the hydrogen gas cloud, leading to its collapse, and the formation of the galaxies and the Titanic superstars of the early universe. These stars, like the Titans, were the parents of the all the later heavy elements, the planets, and the stars to come, including our Sun. We've exercised poetic license here and borrowed the name, so we'll sometimes call these primordial superstars Titans.

All the heavier atoms, such as carbon, oxygen, nitrogen, sulfur, silicon, iron, and so on—the stuff of our rocks, our solid and wet planet, our neighboring planets, our own Sun and neighboring stars, and eventually the stuff of life itself—were created within the gigantic Titans. The heavy elements were baked by nuclear fusion, within their gigantic nuclear furnaces, bound by immense gravitation, deep within the cores of these supermassive stars.[4] Heavy atoms became the raw ingredients of the modern universe, without which there would be no structure. Eventually, by the parentage of the Titans, the planets formed. The specialized conditions on the planets sequentially led to the subtle and gradual evolution of life, and on Earth, of human thought and emotion.

Imagining the early forming of the first stars and galaxies is like traveling to a remote and grandiose place, such as the Alps, the Sierras, or the canyons of the southwestern United States, or viewing the simmering caldera of Yellowstone. The beauty of nature is fully alive and spellbinding in the true scientific story. The saga of the first phases of the universe is common to all living beings that have ever stood, walked, or crawled on Earth or any other planet. The true scientific story of our heritage is richer than any fable, it is more mysterious in its reality, and it may be more comforting to us in its elegant logic. Henceforth we'll banish the fabled gods and immerse ourselves in the natural universe. The story of the real Titans continues as follows.

GÖTTERDÄMMERUNG: TWILIGHT OF THE TITANS

How did the heavy elements become liberated from the cores of the supermassive Titanic stars in which they were formed? Indeed, the nuclear furnace interiors of the Titanic stars eventually poisoned themselves. Filling with iron, the most stable atomic nucleus, they could no longer burn by nuclear fusion. The Titans thus began to collapse. Their hulking bodies, now filled with the newly formed heavy elements, commanded by gravity, now caved inward upon themselves. No longer opposing gravity with the intense radiation of their nuclear engines, a sudden and rapid change occurred deep within their cores. There the atoms of iron, supporting the entire weight of the massive hulk against the collapse by gravity, like the hull of a sinking submarine, gave way and imploded. The iron atoms were squeezed, subject to enormous pressure and density. This instantaneously created a new state of matter, never before present in the universe.

Atoms consist of *electrons* outwardly orbiting the compact *nucleus* that defines the center of the atom. The nucleus is made of *protons* and *neutrons*. When a Titanic star reaches its last stage of collapse, the electrons and protons in its core are squeezed together. A new set of physical processes, normally silently lurking in the background shadows of the everyday world around us, suddenly jumps to the fore. These are called the *weak interactions*, and they quickly convert the squeezed protons and electrons into neutrons, and producing as a by-product an explosive blast of elementary particles called *neutrinos*. The dominant process of the weak interactions that destroys the Titans takes the following form:

$$p^+ + e^- \rightarrow n^0 + v_e.$$

Or, in words, "proton plus electron converts to neutron plus electron-neutrino."

At the instant of the collapse of the core of a Titan, the weak interactions steal the show. The innermost core of the Titan is compressed into a ball of pure neutron matter, extremely compact, perhaps only ten miles in diameter and yet as massive as our Sun, but trillions of times more dense. The neutrinos stream frantically outward from the core. As the neutrinos burst forth, the outer shell of the Titanic star explodes. This marks a *supernova*—the most intense and spectacular explosion to occur in the universe since the big bang.

It is remarkable and ironic that this ferocious "mother of all explosions" involves the lowly neutrino, an elementary particle that seems otherwise to be the most inert and inconspicuous of all particles. Out blast the neutrinos, taking with them all of the outer matter of the star, the newly synthesized elements, producing a brilliant flash of light many millions of times brighter than all of the stars shining within a single galaxy. The outer shell of the body of the Titan, containing all the elements from hydrogen to iron, is blown into space. A dense, spinning neutron star, or perhaps a black hole, the tiny remnant of the pure neutron core of the Titan with a mass greater than that of our own Sun, is left behind.

Over time, the clouds of gas and dust and debris now containing the heavy elements—the cindered remains of the many deceased Titans in their violent fates—accumulated and encircled the galaxies. This gave the galaxies a new and grandiose shape, that of gossamer spirals, with their outreaching and enveloping spiral arms (see the Whirlpool Galaxy in fig. 1). In the outer spirals of the galaxies were born the offspring of the Titans, the second generation of smaller yellowish stars, like our Sun, together with the comets, asteroids, moons, and planets. These were composed of the gas and metallic ashes of the Titans, while the planets were built of the rock made of the elements born in the Titans. These were the true children of the Titans.

The existence of everyday matter, the existence of the planets and the world we inhabit today, the existence of life, and *our very existence* owe to the violent annihilation of these anonymous stars, the primordial Titans that died in the ferocious oblivion of their supernovae, billions of years ago. All of our "everyday matter" baked together within these monstrous conflagrations. This process of heavy-element formation is ongoing throughout the universe, even today. Many Titans exist today, shining with the light of the fusion of pure hydrogen and helium, dwelling within the inner recesses at the centers of galaxies, detonating from time to time. In otherwise dim and distant galaxies millions of light-years away, the supernovae light them up for a moment, flashing in

Figure 1. The Whirlpool Galaxy, M51, shows extraordinarily well-developed spiral arms containing the debris of stellar explosions and the raw material of future star formation. The photo is approximately how our Milky Way galaxy would appear today to a distant observer. (Photo courtesy NASA and the Hubble Heritage team, SCSci/Aura. The image was composed by the Hubble Heritage Team from Hubble archival data of M51 and is superimposed onto ground-based data taken by Travis Rector at the 0.9-meter telescope at the National Science Foundation's Kitt Peak National Observatory in Tucson, Arizona.)

the dark, distant universe like fireflies in the night. And some stars within our own galaxy, and not too distant from Earth, perhaps the unstable and dying η Carinae (EY-ta kar-IN-ee), will one day brighten our own sky with their cataclysmic finales.

EARTH

The Sun, Earth, and our planet's solar-system siblings were born when the universe had attained the age of approximately nine billion years. The solar system condensed like gigantic raindrops in the cloud of dust and debris that was the aftermath of the ancient Titans in the distant arms of the spiral galaxy. A long period of disfiguring bombardment by comets and meteors as well as upheavals of massive earthquakes and volcanic eruptions ensued. The birth and childhood of a planet is not a peaceful passage. By 2½ billion years of age, Earth's continents solidified, and Earth gradually began to host the earliest forms of life. Life beginning on Earth required violent and dynamic conditions to jog chemistry, manipulate sophisticated molecules, and jolt the complex process of reproduction, which defines life, into existence. Life was then in its infancy. Algae proliferated, taking hold in Earth's oceans of seltzer water.

Planet Earth, our blue-green home, cradle to everything that we know, would seem to us now a distant and alien world as it was then. Earth was ending a dark, brutal, and angry childhood. It was beginning to mature and stabilize. Its atmosphere was beginning to acquire oxygen, the waste product of the algae that had breathed and digested the abundant carbon dioxide in the atmosphere and in the oceans. Earth was still highly volcanic and inhospitable to higher forms of life.

Our planet was also *extremely radioactive* two billion years ago. The Titans had produced many elements, including atoms much heavier than iron. These were created during the last violent seconds of the Titan's life—the radioactive debris of a supernova's ferocious nuclear explosion. Uranium was one of the heaviest elements made in the Titanic explosions, and it became incorporated into the original Earth when it formed. Note that "uranium" is named after the grandfather Titan, Ouranos. Uranium is therefore a natural part of Earth's composition.

Today we mine uranium like any other mineral, in deposits where it has been concentrated by the solvent action of water, flowing and diffusing through rock. Among the many practical applications of this heavy, yellowish metal, it has been used to make nuclear reactors and nuclear weapons. Scientists define *uranium* to be any atom with a nucleus that contains 92 protons. However, the number of neutrons in the nucleus can vary, giving rise to different *isotopes* of uranium. Today the uranium found in mines is mostly of the ^{238}U form (read this as "U-238") with a tiny fraction of the ^{235}U form ("U-235") variety. The number 235 refers to the *total number of neutrons plus protons* in the nucleus; hence ^{235}U has $235 - 92 = 143$ neutrons. The isotope 238 of uranium therefore has three more neutrons in its nucleus than the 235 form. When uranium ore is mined today, it contains 99.3 percent of the ^{238}U form and a mere 0.7 percent of the ^{235}U form.

The process of "splitting apart" of atomic nuclei is called *fission*. Nuclear fission can occur only in the heaviest elements, those much heavier than iron, and in the process of fission of a heavy nucleus, a large quantity of energy is released. It is this release of energy that can drive a nuclear reactor or atomic bomb through a *sustained or runaway chain reaction*. To construct a nuclear reactor or nuclear weapon, we must *enrich* the ^{238}U form by increasing the fraction of the ^{235}U in the mixture. In enriched uranium, the fission of a single nucleus produces a few rogue neutrons and lighter "daughter nuclei," which become new atoms. The rogue neutrons roam around until they strike another uranium nucleus, which in turn triggers that nucleus to undergo fission, producing more daughter nuclei, more rogue neutrons, more energy, and so on.

With only a small amount of fissionable material, a sustained chain reaction does not occur. Most of the rogue neutrons simply cross the boundary of the material before hitting another nucleus. If, however, *enough* enriched uranium is concentrated together, to form a *critical mass*, then the chain reaction becomes sustained. With a *supercritical mass*, the chain reaction accelerates and "runs away." The uranium heats up to enormous temperatures, ultimately melting, bubbling, churning, and flowing. If, however, it is simultaneously compressed by a conventional explosive, the supercritical mass will explode—the principle of the atomic (fission) bomb. A slow, self-sustaining nuclear reaction occurs when the mixture contains about 3 percent or more ^{235}U and 97 percent ^{238}U. *Weapons-grade uranium* involves a significantly higher fraction of ^{235}U, typically greater than 90 percent.

When the many Titans of our young galaxy exploded, roughly equal amounts of these two different isotopes of uranium were produced and hurled out with the debris that made the spirals of our galaxy. This debris became incorporated into our planet Earth. Why, then, is the ^{235}U isotope such a tiny fraction of the uranium we find in mines on Earth today? The reason is that the atomic nucleus of ^{235}U is more unstable, spontaneously decaying at a faster rate, than the nucleus of ^{238}U. Physicists find that the *half-life* of ^{235}U is about 700 million years, or roughly one-sixth of Earth's present age. This means that one ounce of ^{235}U today will be reduced to one-half ounce in 700 million years. The other one-half ounce will be in the form of other, lighter atoms that are the by-product of the decay process. The half-life of ^{238}U, on the other hand, is about 4.5 billion years, much longer than ^{235}U and about equal to the age of Earth itself. Therefore, the older Earth gets, the smaller and smaller the fraction of ^{235}U becomes relative to the longer-living ^{238}U. Over the age of Earth, the longer-living ^{238}U has come to dominate the planet's abundance of uranium.

Two billion years ago, however, the abundance of ^{235}U was therefore much larger than it is today. In fact, it exceeded 3 percent of the ^{238}U form. Thus *enriched uranium was then a naturally occurring substance on Earth.* Since enriched uranium was naturally present in the young Earth, our metaphorical Mother Gaia did something remarkable: *she made her own nuclear reactors*! These nuclear reactors were created in dense mineral deposits by the natural concentration of uranium into large shallow veins, by the flow of water and diffusion into the cracks and fissures within rocks. Nature's own nuclear reactors were shapeless and amorphous blobs, like the molten core of a modern nuclear power plant disaster—naturally occurring Chernobyls, so to speak. They broiled and churned within the embracing rocks, spewing molten radioactive waste, blowing radioactive steam and gas through geysers and roaring vents. They ate through their fission fuel while poisoning themselves in their own radioactive wastes. Then the wastes diffused and boiled or decayed away, and the reactors restarted themselves again. So it was, these natural reactors repeated the process, turning on and off, again and again, over a period of millions of years. Finally, the natural reactors exhausted their enriched-uranium fuel and quietly died.

OKLO

The remains of one of seventeen ancient, naturally occurring nuclear reactors were discovered in 1971 amid a uranium ore deposit, in the village of Oklo (pronounced "oak-low"), Gabon, West Africa. While active, the Oklo natural reactors had generated radioactive waste products identical to those of modern nuclear reactors at power plants. Only one of the original seventeen sites at Oklo is particularly conspicuous now, since fourteen of the others had been mined out prior to the discovery in 1972. Two of the ancient reactors remain to be explored.[5]

The remains of this fossil nuclear reactor are visible in an underground tunnel wall. They appear as a seemingly unnatural, light-yellow-colored rock that is composed mostly of uranium oxide, with streaks of shimmering quartz glass. The quartz is crystallized silicon, produced from the bath of the superheated underground waters as they circulated sand through the reactor's core during and after its lifetime. The Oklo reactors produced all the usual fission by-products, such as ^{239}Pu (plutonium-239), which is a horrifically toxic and highly radioactive element, also used for weapons. The plutonium burned in its own fission process, together with the enriched uranium. Because plutonium has a relatively short half-life of a mere twenty-four thousand years, there was essentially no plutonium present in the debris cloud when Earth formed, proving that the Oklo reactors were indeed nuclear reactors and had produced the plutonium themselves.

The Oklo nuclear reactors are a stunning natural phenomenon. At the time that the Oklo nuclear reactors were spontaneously burning their fission fuels, the universe was about 15 percent younger than it is today. This leads us to reflect upon the hypothesis of the eternal constancy of nature itself. Might the universe and its laws of nature two billion years ago have been slightly different than they are today? Was gravity slightly different, weaker or stronger, then? Were the electromagnetic forces of nature the same? Were the laws that govern the nuclear processes exactly the same in the earlier universe as they are today?

The Oklo nuclear reactors provide a remarkable and sensitive window on physics, as well as the basic laws of nature of the world, as they were two billion years ago. All nuclear reactors create various rare elements as by-products of their nuclear reactions. These involve the extreme processes that can happen only in stars or nuclear reactors, processes that are delicately sensitive to the exact laws of nature. Prior to modern nuclear reactors, this was the only time that these elements were synthesized on Earth. One of these nuclear processes led to the synthesis of the particular rare element called *samarium*, with the chemical symbol Sm.

Samarium was discovered in Paris in 1879 by Frenchman P.-E. Lecoq de Boisbaudran. This lovely shiny, silvery-colored, nontoxic metal has a brilliant sheen. Most samarium found on Earth is primordial, having been produced by the Titans. It is usually found within geological formations in several minerals and can be chemically separated from the other heavy atoms that usually accompany it. It is employed in manufacturing the bright lights used for movie projectors and in certain kinds of lasers, as well as in the construction of nuclear reactors themselves.

From Oklo we indeed learn something subtle, yet profound, in nature's nuclear engineering feat—the abundance of samarium produced in the natural nuclear reactors at Oklo of two billion years ago *is exactly what we would expect it to be today*! Why is this so remarkable? Indeed, we know that the production of this by-product of nuclear fission is extremely sensitive to the complex physics occurring within nuclear reactors. If there were tiny differences between the basic laws of physics back in the time that the Oklo reactors functioned, two billion years ago, then absolutely no samarium could have been produced. Therefore, Oklo, together with its samarium by-product, showing up in the correct abundance, tells us that *the universe had to have the same laws of physics two billion years ago as it does today*. In fact, from the measured abundance of samarium at Oklo, scientists can infer that the relevant laws of physics cannot be changing in time by more than 1/1,000,000, one part in a million, over the age of the entire universe.[6]

STABILITY OF THE LAWS OF PHYSICS

Laws of physics that somehow change in time are a bizarre and unsettling thing to contemplate. How, indeed, might the laws of nature have been different in the earlier universe of two billion years ago to affect the way samarium is produced in a nuclear reactor? It turns out that *a very tiny shift in the mass of the atomic nucleus* of samarium would have been enough to *block its formation* altogether in the Oklo nuclear reactors. Theoretically we could imagine that this might have happened in many different ways, but only if the laws of nature were somehow different back then. For example, if the quantitative value of the unit of the electric charge of the electron or proton were slightly different two billion years ago, that tiny difference would have affected the electromagnetic interaction between the protons in the nucleus. This would have slightly changed the mass of the nucleus of a samarium atom by a corresponding amount. Through an analysis of the abundance of the Oklo samarium, however, scientists have calculated that the magnitude of the electric charge could not have changed by more than 1 part in 10,000,000 (one part in ten million, or 10^{-7}) at the time Oklo was burning uranium. This means that the value of the electric charge cannot be changing by more than 1/100,000,000,000,000,000 (one part in one hundred million billion, or 10^{-17}) per year today! This is a revealing and somewhat reassuring discovery about the constancy of the laws of physics through time.

Oklo is not alone. There are many other indicators of the stability of the laws of physics through time. Astronomers can peer through telescopes at distant galaxies and see that the same physical processes are at work in those long ago and faraway worlds as are occuring here in our laboratories on Earth today. The abundance of certain elements in meteorites tells us that other very sensitive processes are the same today as they were billions of years ago. In the 1970s, the Viking mission, sent to Mars by NASA, allowed a precise measurement of the force of gravity, which determined that it, too, is not changing through time. Taken together, all of the experimental evidence suggests a reasonable hypothesis about the laws of nature: *the laws of physics are constant and are not changing through time.*

Eternal constancy of the laws of physics is a *symmetry*. What we see as we look back in time, or we peer through telescopes out into space, or we look through our powerful microscopes (particle accelerators), is the same system of laws of physics governing the whole universe at all times and at all places. These are the basic symmetries of the structure of our universe and its contents and, at a deeper level, the symmetries of the laws that govern the universe themselves. Indeed, the symmetries we uncover are the basic principles that define our laws of nature and the laws of physics, hence those that control our universe. And, as we will now see, constancy of the laws of physics has immediate consequences for our everyday existence.

chapter 2

TIME AND ENERGY

Energy is eternal delight.
　　　　　　　　—William Blake, *The Marriage of Heaven and Hell*

IT CAN'T HAPPEN HERE

The Acme Power Company does not exist, and has never existed, to our knowledge. Any resemblance of this power company to any other power company, operating or not, past or present, living or dead, its managers or investors, real or fictional, in jail or out on bond, is a mere coincidence. We've invented the Acme Power Company to make a point about physics.

Countless companies like Acme have no doubt existed throughout history. Unfortunately, they promise something for nothing and untold wealth to the investor who gets in on the ground floor. Not that we mean to impugn or imply malicious mischief on the part of the founding fathers of the Acme Power Company. The whole incident was an honest mistake—at first. However, as things got going, ever so imperceptibly, they became unstoppable and gathered momentum. Many analysts, bankers, promoters, and high-minded and well-meaning politicians entered the fray, having a vested interest in its promise. Before long the Acme Power Company was proclaimed to be a success, whether it was or was not, since the alternative would be unthinkable. In the end, however, the laws of physics would rule the day.

The Acme Power Company was formed by a small group of wealthy investors, having heard the claims of an obscure inventor who had found a "new way to generate electrical power." The inventor had discovered in his basement laboratory that the laws of physics are changing in time. He had noticed variations in the force of gravity throughout the course of a week, particularly on Tuesday mornings. The force of gravity was observed to be consistently weaker every Tuesday at exactly 10:00 AM. The business plan of the investors was to extract energy from the changing force of gravity because of the strange "Tuesday phenomenon." Since on Tuesdays the force of gravity was weaker on the surface of Earth than on any other day of the week, a large mass, composed of any substance, could be hoisted into the air with less cost in energy than at any other time during the week. Then, later in the week, the mass could be released, and would return more net energy than was consumed in hoisting them.

A brief technical aside is in order. The force of gravity at the surface of the Earth is measured in terms of g, the *rate of acceleration* that any object, such as a rock, experiences (neglecting air resistance) when it is dropped, perhaps from the Leaning Tower of Pisa. The *force* that an object of *mass*, m, experiences on Earth's surface, due to the pull of gravity, is simply the product of mass times the acceleration of gravity, or mg. As every high school student learns in her physics class, the acceleration due to gravity on Earth, g, is about ten "units" in the *meter-kilogram-second* system of measurement.[1] That is to say, g is roughly 10 meters per second-squared, or equivalently, 10 m/s^2 (this is equivalent to 32 feet per second-squared in the English system of measurement[2]). This means that, after falling for one second, and neglecting air resistance, any mass will have a speed of 10 meters per second (32 feet per second[3]). In short, increasing the force of gravity would increase the value of g.

According to Acme's inventor, every Tuesday at 10:00 AM, for a few minutes, g was significantly less than on every other day of the week. We therefore would all weigh a little less on Tuesday mornings at 10:00 AM. This effect was measured on the Acme patented g-meter, built by the inventor in his basement lab, who claimed it was a very accurate way to measure g.

The Acme Power Company, after floating an initial public offering of a million shares of stock, purchased a large water tower, a reservoir, and a water turbine electrical generator that could be run in reverse as a pump. The water tower, which was high above the ground, could hold a large amount (or mass) of water. Therefore, from a formula known to every high school physics student, the *total energy* required to pump the mass, m, of water into the water tower, at height h above the ground, is the product m times g times h, or mgh.

On Tuesdays at 10:00 AM, the Acme g-meter indicated that gravity had become weaker, or g had become smaller, becoming only nine units in the meter-kilogram-second system of units. So, the water tower was quickly pumped full of water from the reservoir (see fig. 2). The company got the energy to pump the water up the tower from the power lines. The water was then allowed to sit in the water tower overnight.

On Wednesdays the Acme g-meter showed that gravity had returned to its original strength. That is, g had returned to the larger, standard value of ten units in the meter-kilogram-second system. A valve was opened and the water was allowed to flow down from the tower through a system of pipes, and passing through the Acme turbine electrical generator, back into the reservoir. The gravitational potential energy of the water pumped up to the height of the water tower was now recovered and converted back into useful electrical energy. But g was now larger (10 units) than it was on Tuesday (9 units), and the energy extracted from the water as it flowed back down was *greater* than the original cost of energy in pumping the water up. The Acme Power Company therefore claimed to get a *net excess amount of energy* out of the system equal to the product m times h times (g_{Wed} minus g_{Tues}), or $mh(g_{Wed} - g_{Tues})$.

Now, energy is a commodity that has a dollar value assigned by the market to each unit of energy. This recovered energy could repay the cost of the energy used to pump the water into the tower, and the extra remaining energy could be sold as pure profit by feeding it into the power grid. This system could therefore provide electricity for all the nearby towns and their citizens. The Acme Power Company was producing net energy for free from the time variation of gravity. The Acme Power Company had in essence constructed a so-called free-energy machine that could run indefinitely, producing more energy than it consumed, and all for free![4]

As rumors swept Wall Street of this breakthrough, the stock in the Acme Power Company soared higher and higher. The managers of the

Figure 2. The Acme Power Company test facility is shown, consisting of a water tower of height h, a turbine generator of 100 percent efficiency, and a storage pond from which the water of mass m is pumped up into the tower, by running the turbine generator in reverse. The Acme "g-meter," which measures the rate of acceleration of gravity on Earth's surface, g, is shown on the lower left. (Illustration by CTH.)

company stated, "It is only a matter of time until the first Acme systems will be up and running, sending energy to all of the subscriber communities, netting millions for the investors." Many orphans and widows had their nest eggs invested by bankers and stockbrokers in this "no-brainer" stock. It had become the overnight darling of Wall Street.

A suspicious auditor, however, requested that the Stocks and Change Commission (SCC) hire an independent laboratory to do a test on the Acme Power Company's system. In particular, the g-meter, which had revealed that the laws of gravity were dependent upon time, was to be subject to a number of careful and precise tests. In June the g-meter was procured by the SCC authorities and turned over to the Universal Testing Laboratory (UTL). It was announced that the results of the test would be released sometime within the month of October. The stock trading became frothy toward the end of the summer as eager investors retreated to and advanced from the sidelines, waiting for UTL's results, and the news that would confirm the great breakthrough of the Acme Power Company and its obscure yet daring inventor.

Finally the month of October arrived. Shareholders often become nervous in October. As the great investment adviser Pudd'nhead Wilson once observed, "October . . . this is one of the peculiarly dangerous months in which to speculate in stocks—the others are July, January, September, April, November, May, March, June, December, August, and February."[5] The eve of the long-awaited laboratory test on the g-meter arrived, and the Acme Power Company stock momentarily surged down at the close of trading. A scurrilous rumor had swept the floor of the exchange that the g-meter's inventor had disappeared, having hopped the red-eye flight to somewhere in Eastern Europe the previous morning.

Just before the start of trading the following day, the results of the analysis by the UTL were to be announced. The "street" waited with bated breath, and a drumroll could almost be heard as the moment approached. Finally the announcement by the officials of the UTL was read, and the story went out on the wire: the tests had revealed that the Acme Power Company's famous g-meter was indeed reading a lower value at 10:00 AM on Tuesdays—but this was due to a *faulty design*!

A careful analysis had revealed that the air-raid sirens in the neighboring towns were tested every Tuesday at exactly 10:00 AM, which caused an acoustic vibration in the sensitive circuitry of the machine and a slight reduction in voltage to the g-meter readout. This gave a false reading of the physical quantity g, which was misinterpreted as a reduced force of gravity. Upon correcting for this systematic error in the g-meter, the testers found that there was absolutely *no change* in the value of *g* on Tuesdays. They asserted that the laws of physics don't appear by this, or any other, known experiment to be changing in time. In its report, the Universal Testing Laboratory cited the famous discovery of the Oklo natural nuclear reactor, stating, "Since the value of the unit of electric charge does not change by as much as one part in ten million over the lifetime of the universe, based upon the findings at Oklo, it is quite unlikely that gravity varies in strength by as much as 10 percent during the course of the business week. Although *g* is a physically different quantity than electric charge, the Oklo results generally confirm, strongly, the constancy of the laws of physics to a precision far exceeding the signal registered in the faulty *g*-meter."

Hence, the Acme Power Company's elaborate system produced *no excess electrical energy*. In fact, it only lost energy into its various waste forms, such as heat, mechanical vibration, noise, and so forth, due to the inherent inefficiencies found in any mechanical or electrical system. The laws of physics do not change with time but are constant.

The trading in the stock of the Acme Power Company was immediately suspended by the SCC. For days not a single share of stock (legally) changed hands. Days turned into weeks, and weeks into months. When it finally resumed trading, the darling of Wall Street, once a three-digit high-flyer and prestigious cover story of *Blurbes Magazine*, was now a penny stock. Later, when the SCC launched a criminal investigation, it was found that the poor inventor of the *g*-meter had been initially fooled by this effect in his basement lab but later told the investment bankers about the problem after he had discovered it. But by then, the Acme Power Company stock was soaring higher and higher, so it was decided, by someone somewhere in a smoke-filled room, not to spoil a good thing.

It was later revealed that the CEOs, CFOs, presidents, board members—both current and former—and some very large investors in the Acme Power Company had actually sold all their shares some months before UTL's results were announced (all for completely legitimate reasons, of course). They had, however, continually assured the investors and employees who had their retirement plans invested in company stock that "all is well—we'll be making lots of free electricity in the next quarter. Be patient!" The officers in the company completely denied any knowledge of this *g*-meter problem. A financial clerk in the back office, however, went to jail for misfiling a memo in which the *g*-meter was discussed at a board meeting.

That was the end of the Acme Power Company.

As we said at the beginning, the story of the Acme Power Company is only a fable. One might think that this story is absurd and that no self-respecting investor, in reality, would ever be bilked into buying stock in such an inane scheme. But, in fact, countless perpetual-motion and free-energy machines have been proposed and have bilked investors over the centuries. Many patents have been awarded for such creations, even into the late twentieth century. The particular setups vary considerably. In the early nineteenth century, these systems may have employed water that fell from buckets on conveyor belts, pushing one bucket down, which in turn raised other buckets that dropped more water into other, lower buckets, and so on. Or the system may have involved pumped water driving pistons that pumped more water that drove the pistons, and so forth.

More modern perpetual-motion or free-energy systems often involve an apparently more sophisticated degree of technology. For example, they may involve the *electrolysis* of water, in which one breaks apart ordinary tap water, good old H_2O, by passing an electrical current through it, thus converting it into its constituent gases of H_2 and O_2. The H_2 and O_2 can then be chemically combined (burned) in an internal combustion engine, which gives back energy, plus H_2O, ordinary water, as exhaust. Electrolysis certainly works and is often demonstrated in a high school chemistry class. Unfortunately, however, electrolysis has been *misunderstood* by the investment community, which has sometimes been led to believe that burning the resulting H_2 and O_2 produces *more energy than the original electrolysis of the water consumed.* This is definitely not true! This process has been claimed from time to time to be the source of limitless energy that can power automobiles and cleanly generate electricity forever after.

In the 1970s one such company attracted the attention of investors, and one morning its stock price rocketed upward at the opening of trading. One of the authors (CTH), then a graduate student at Caltech, arrived at the physics building to find Richard P. Feynman, the world's greatest theoretical physicist, somewhat amused by the situation and asking, "How can I short all of this company's stock?" By this time, the stock's trading had already been halted. However, "put options" were apparently still being traded (which is an approximately equivalent way of shorting a stock). Feynman and the student had lunch and discussed how to place a position in puts. Upon hearing that it required the completion of special forms and permission from the brokers, Feynman abruptly exclaimed, "This is a ridiculous idea and a waste of time—I'm going back to my office to do some physics." Remarkably, the trading in the company's stock later resumed, yet the stock price did not crash down into pennies per share as one may have expected—in fact, it never returned to its original lows! Somehow, the believers doubted the counterclaims that the perpetual-motion free-energy mechanism couldn't possibly work. The put options expired worthless. We can therefore state that the stock market is *not*, evidently, ruled by the laws of physics.

It's easy to get confused by this—if you already have a source of energy, such as windmills or nuclear power, you *can* readily convert water to pure hydrogen gas (and oxygen) and even use the hydrogen as a fuel. However, you are then *using up the energy you started with*, not creating a net amount of free energy from nowhere. Like the Acme Power Company, any free-energy water-to-hydrogen conversion process would have necessarily constituted a perpetual-motion machine. It, too, would have produced net excess energy for us from nowhere, which we could convert into cash. This could work only if the laws of physics were somehow changing in time. To be a net energy source would require that the energy invested in breaking apart water molecules be somehow less than the energy gotten back by burning the hydrogen and oxygen, recovering the water molecules. Hence, the properties of the initial water molecules would have to be different somehow from the properties of the final water molecules. However, water molecules are comparatively simple physical systems. Any water molecules that were born in the early universe—by-products of the explosions of the Titans—have exactly the same physical properties that water molecules have today. Their properties do not change with time. We cannot extract net energy by breaking them apart, then recombining them into their original form.

Indeed, it may (or may not) be a good idea, for future energy policy, to create hydrogen fuels for automobiles and other things by electrolyzing water from some other clean, centralized source of energy. The burning of hydrogen and oxygen is relatively safe, clean, and efficient and does not pollute the air with carbon compounds. It appears to be environmentally friendly. Its overall environmental impact may not yet, however, be fully understood. It would require a vast change in our energy infrastructure. And we will need some new raw source of energy to do this—we won't gain net energy from this process. In fact, we will lose energy, since the overall process cannot be 100 percent efficient. Whether this is in our future or not depends upon the problems that are encountered in the first large-scale operations. So far, however, it looks promising.

If energy could be produced from nowhere, or vanish into nothingness, we would say that energy is not conserved. The total energy that the Acme Power Company used to pump the water into the water tower was supposedly less than the energy gotten back by releasing it. Thus energy would not have been conserved—net energy would have been created from nothing. However, in every experiment that has ever been conducted to measure such effects we have always discovered that *the total energy we begin with equals the total energy we end up with*. Thus energy is conserved in nature. Countless scientists have performed numerous experiments like this to verify that the total initial energy was indeed the same as the total final energy in any physical process.

One of the reasons it is so easy to become confused about energy conservation and to ruminate about perpetual-motion or free-energy machines is that energy is hard to keep track of. Many things in our everyday lives are not conserved: the number of living organisms on planet Earth and the total value of the stock market are two examples. Energy also takes many different forms. It is quite apparent in a moving object (kinetic energy) but less so in an object sitting at rest on top of a mountain (potential energy, which can be converted to kinetic energy as the object falls). Energy is generally lost in physical processes that convert it into waste forms such as heat and sound. It can be lost in the deforming of materials, creating dents and crumples, which changes and rearranges the molecules in the material. Energy can be absorbed (or released) in the form of chemical energy, changing the physical state of matter from solid to liquid or from liquid to gas. Energy can stream out of a system carried by light and other forms of radiation. A system, such as a large star that has run out of fuel, can shrink, reducing its gravitational potential energy, which is converted and radiated as light, until the energy is exhausted and the star becomes finally a white dwarf or even a black hole. Indeed, it took physicists, chemists, and biologists a long time to understand that the principle of conservation of energy is exact and omnipotent. It governs everything. Even life-forms are governed by energy conservation. There is no special form of energy reserved for living things—all energy can be measured by the same units throughout the entire universe. If you could do all of the detailed bookkeeping and keep track of all forms of energy, you would find that energy is always conserved in any process whatsoever.

What we have seen with the Acme Power Company is that, if the laws of physics were changing in time, then one of the most important principles of physics, the principle of energy conservation, would cease to be true. If the forces of nature are different at one time than at another, then the amount of energy invested in a physical process would be different than the amount of energy invested in the same process at a later time. However, we've learned from Oklo, as we've learned from many other diverse observations, that the laws of physics are *not* changing through time—*over time scales almost equal to the age of the universe*. Thus the result of any particular physics experiment that we perform tomorrow, or yesterday, or ten seconds ago, or ten billion years ago, or a thousand billion years in the future, will produce the same results. The laws of physics—and thus all the correct equations in physics—are the same at any time in the history of the universe. This is an experimental fact. The laws of physics are steadfast and eternal.

We have just glimpsed one of the most important relationships in nature: energy conservation is associated with the fact that the laws of physics do not change in time! This is the first example of a phenomenon of a more general and profound significance known as Noether's theorem. The key point is that the unchangeability, or *invariance*, of the laws of physics is a *continuous symmetry* of the laws of physics. Noether's theorem says that for every continuous symmetry of the laws of nature, there is a conserved quantity.

BUT WHAT IS ENERGY?

Many of the challenges of paramount importance that are facing our civilization today revolve around the subject of energy. The reason for this is simple: energy is the primary commodity that we consume. Thus the causes of many wars and conflicts in which we find ourselves continually immersed have a basis in the need for an abundant and convenient form of energy. In modern times, this has been oil. Energy is the key to our economy and to our future, as well as to our political power and authority. The proper use of energy is, however, fundamental to the fate of our environment. One might say that the two most urgent issues confronting the human species are world overpopulation and energy policy, perhaps in that order, and they are inextricably intertwined. These are the two most difficult issues from the point of view of serious public-policy making— there's no evidence that they can be ameliorated by any nonviolent political system humans have yet designed.

In addition, energy is a very poorly understood concept. Sometimes we hear phrases like, "He has so much psychic energy," or, "The body's energy flows as the life quantum through the focal points of the mystical crystals, upward through the tip of the pyramid," etc. These phrases are referring to something that has nothing to do with energy as physicists understand it. Usually these are bogus notions, or metaphors at best. Unfortunately, energy has acquired, in many quarters, a kind of mystical interpretation, which many people accept.

We also might hear someone say, "A month after the surgery she finally got her usual energy back," or, "He is lacking in mental energy," which is a kind of poetic description of strength or animation, as in vim and vigor or intellectual ability. This usage is fine as a figure of speech. It is not, however, what physicists have in mind as a valid physical definition of energy, either. Although energy has several meanings in our language, it has only one very precise meaning in physics.

Nonetheless, although most any physicist can readily define energy of a particular type, devising a general definition isn't a simple task. In high school physics books, *energy* is defined as "the ability to do work." Great! But this requires a precise definition of *work*. In physics, definitions must be crystal clear and unambiguous, the ultimate test being the ability to write a mathematical equation. The trouble here is that *work* in physics is a bit complicated, being "the force vector dotted into the displacement vector of an object." So, for the moment, trust us that energy has a precise definition for all of its various forms, but let's briefly consider a few specific forms of energy (perhaps to draw some sympathy for our problem of providing a more precise and universal definition).

Kinetic energy is energy of motion, and it depends upon the mass and the speed of a moving object. It requires energy to make a massive object move, requiring more energy the more mass that the object has and the more speed that we desire of the object (shortly we'll consider the motion of an automobile). When a substance contains molecules or atoms that have a lot of kinetic energy, involving fast motion in random directions within the substance, we say that the substance is "hot." When the kinetic energy of the molecules or atoms is small, the substance is "cold."

Potential energy is energy stored in an object or system that is ready upon its release to make other objects move. For example, a compressed spring has potential energy that can launch a toy dart from a child's rubber-tipped dart gun, help to hoist open a garage door, or run an old-fashioned wind-up watch for days. This energy in a spring is actually the *energy of deformation* of the lattice of iron alloy (steel) atoms within the material as they are twisted slightly away from their normal relaxed pose. There can be many forms of potential energy. For example, a bank of snow sitting atop a mountain has gravitational potential energy, ready at an instant to fall and convert to kinetic energy of motion. Gasoline and other fuels contain chemical potential energy waiting to be released by the chemical reaction of oxidation (burning).

Chemical energy is created (or consumed) as various substances can undergo a vast array of chemical reactions that produce or consume energy. The precise form that the chemical energy takes depends upon the reaction. One rather common example is the burning of coal, oil, wood, or other substances that have a high percentage of carbon. Burning is the act of combining carbon with oxygen (a gas that is conveniently and generously supplied by the atmosphere). The basic reaction is $C + O_2 \rightarrow CO_2 + Q$. Here Q is a symbol for energy, which includes particles of light (known as photons, or equivalently, the particles that make up electromagnetic radiation), and the rapid motion (kinetic energy) of the resulting molecules after burning. In other words, carbon combines with oxygen to produce carbon dioxide plus energy.

The high speeds of the resulting molecules after burning is random and is called *thermal energy*. In a wood fireplace, the rapidly moving molecules, products of the combustion, collide with other molecules, such as the surrounding air, giving them kinetic energy, which propagates the heat outward into the room in a process called *convection*. Photons stream out into the room as well, as *thermal radiation*, producing radiant heat. The pleasurable sensation of a warm fire in a fireplace is nothing more than a bath of faster-moving air molecules and photons.

Electrical energy is yet another form. In its simplest form, it is just the kinetic energy of the flow of electrons (an electric current) through a wire, through certain liquids, or in free space, as in a vacuum tube (such as a cathode ray tube, a television picture tube) or in an electron particle accelerator. If the wire has a large *electrical resistance*, then the electrons collide with the atoms in the wire, losing their energy, causing the wire atoms to move. The wire thus becomes hot, as in a toaster or an electric oven broiler. This is called electrical resistance, and it leads to the loss of electrical energy. The tricky thing about the bookkeeping of electrical energy, however, is that an electron can "radiate," or *emit*, a photon, the particle of light, through the process $e \rightarrow e' + \gamma$. Here the electron, e', after the emission of the photon, γ, has less energy than the electron, e, before the emission, and the emitted photon has been created, carrying off a certain amount of energy.[6] This can also go in reverse, $\gamma + e \rightarrow e'$, where the initial electron, e, *absorbs* the photon and the final electron, e', has gained the photon's energy. This is the fundamental process in nature that defines electromagnetism, and we will ultimately see that it comes from fundamental symmetries of electrons and photons. Photons can be stored in an "electromagnetic field," as a kind of photon soup that itself contains the energy of the photons. So, energy in the basic processes of electricity and magnetism is difficult to keep track of, involving continual swapping back and forth between electrons and photons. Chemical energy, when examined microscopically, is really electrical energy within atoms and molecules.

Human activity—indeed, human life—always requires the consumption of energy. Of course, if we had essentially infinite energy resources of the right kind, when Earth became too overpopulated we could leave and go to other planets and make them habitable. We could hollow out asteroids and live inside of them like cosmic cave dwellers. Or we could turn Mars into a beautiful planet like Earth. We could even drag Mars into a more convenient orbit, closer to the Sun, and drop comets (large ice cubes) on the planet to give it a nice ocean. We could create an atmosphere for Mars through various chemical processes, driven perhaps by nuclear fusion. This would be the ultimate "habitat for humanity" project. All of this is just a question of energy, know-how, and time. Therefore, in principle, the human population could become arbitrarily large if we had the energy required to make other planets habitable.

However, we don't have this capability at present, or in the foreseeable future. As we overpopulate planet Earth, we therefore create nasty problems associated with our need for energy. At present we predominantly depend upon the burning of carbon for our energy. The burning of carbon produces the products of combustion, such as carbon dioxide and other waste gases, as emitted in the burning of gasoline by automobiles. These carbon-containing gases allow the higher-energy photons of visible sunlight to pass uninhibited through the atmosphere. But the copious lower-energy photons of thermal radiation, produced when sunlight heats the surface of Earth, are absorbed by these so-called greenhouse gases. This traps energy, which heats the planet. The activities of six billion people (approaching ten billion within this century) all producing these gases through combustion, dispersing them into the atmosphere, serves to create a menacing environmental effect of global climate change. In burning carbon-based fossil fuels, therefore, we are potentially causing a global climate change of incalculable severity. We are also setting ourselves up for an ultimate letdown—the lack of available energy when all the carbon-based fossil fuel is gone. This may happen sometime in the present century.[7]

As we have indicated, energy is a precisely defined concept in physics. It is a useful concept because it is conserved in all processes. If we have a large box inside of which all possible things can happen, such as springs compressing and expanding, various bodies falling and bouncing, water flowing, chemicals reacting, objects burning, atomic nuclei disintegrating, and so forth, through all of this there is one number that stays the same—the *total energy*.

As a simple example of kinetic energy, consider a familiar moving object, such as an automobile. Suppose the automobile has a mass typical of compact automobiles, about 1,000 kilograms.[8] We'll assume that the automobile is traveling down the highway at a speed of 60 miles per hour, which is approximately 30 meters per second.[9] Physicists then compute that this automobile has a *kinetic energy*, or an energy of motion, equal to 450,000 energy units. They get this number by multiplying ½ times the mass in kilograms times the speed in meters per second times the speed in meters per second.[10] The answer comes out in the particular energy units called *joules*, when we use the meter-kilogram-second, or MKS, system of measurement. The energy units in the MKS system, joules, are named after James Prescott Joule, a nineteenth-century physicist who spent a great deal of time measuring and studying energy, especially when heat or thermodynamics was involved. (Joule also invented electrical arc-welding.) The statement that our automobile has a kinetic energy of 450,000 joules is a scientifically precise statement about the motion of the car and its kinetic energy.

For comparison, consider a completely different, and somewhat more bizarre, physical system—the motion of a pulse of protons in the Fermilab Tevatron, which is currently the world's highest-energy particle accelerator. One pulse in the Tevatron may contain about three trillion protons, about the number of atoms in a single living cell. The pulse is accelerated until it travels at 99.9995 percent of the speed of light. We cannot use the simple formula we just used above for the automobile to compute the energy of the pulse of protons, because that formula comes from the physics of Galileo and Newton, called "classical physics," and its validity breaks down when things are traveling near the speed of light.

Fortunately scientists know what to do in this case—they seize upon Einstein's special theory of relativity, and from this they can correctly compute the energy of the pulse of protons.

Therefore even something as far removed from our everyday experience as a pulse of protons traveling near the speed of light has a definite value for its energy. The pulse we have described, remarkably, also has an energy (using Einstein's theory) of about 450,000 joules, the same energy as the kinetic energy of an automobile traveling at 60 miles per hour down the highway! Energy is a well-defined physical quantity that describes everything in the universe, and it always has a precise meaning. Energy is conserved in every physical process, and if we had perfect energy conversion efficiency available to us, we could convert the energy in the Tevatron pulse to make an automobile accelerate to 60 mph, or vice versa.

The example of the Acme Power Company showed that energy is a commodity. It is not created or destroyed; it can only be converted from one form to another. However, this conversion process is always inherently inefficient. In fact, the entire field of thermodynamics, a great subdiscipline of the science of physics, was developed to deal with the issue of energy conservation and the inefficiencies inherent in energy conversion. An *engine* converts one form of energy, usually chemical or thermal energy, into another form, usually kinetic energy, in order to move things. Engines never create net excess energy and always lose some energy in the process due to inefficiency. Physicists have proven that there can never be a perfectly 100 percent efficient engine (see, e.g., the discussion of Carnot efficiency and the history of thermodynamics in note 7). The Acme Power Company claimed to have an engine that was 110 percent efficient—supposedly producing more energy than it consumed.

The time rate at which energy is produced, consumed, or converted is called *power*. One can think of power as a kind of speed if one thinks of energy as a kind of distance. If you want to take a trip somewhere, you must travel a certain distance. How fast you do this depends upon your speed. The greater the speed, the shorter the time for the trip. Likewise, you may want to consume a certain amount of energy to perform a task, such as mowing the lawn. How fast you perform the task determines the power you require, the time rate at which you consume the energy. The more power you consume, the shorter the time the task will take. Note that power is not a fixed or conserved quantity, because we can speed up or slow down the rate at which we perform the task. The total energy, on the other hand, is fixed, like the total distance traveled for a given trip.

So, how much power is the typical 1,000-kilogram automobile consuming as it travels down the highway? One way to measure this is to get a car going 30 meters per second (60 miles per hour) on the open highway. (Be sure that you are the only one on the road and do this experiment very carefully!) Then just take your foot off the gas and let the car coast—carefully, far away from any other traffic! Measure how long, in seconds, it takes for it to slow down to 25 meters per second (50 miles per hour). At that slower speed the energy is ½ times 1,000 kilograms times 25 meters per second times 25 meters per second, which is 312,500 joules. Therefore the car has lost 450,000 joules less 312,500 joules = 137,500 joules of energy of motion. If the car slowed down in a period of 10 seconds, then it lost energy at the time rate of 137,500 joules/10 seconds = 13,750 joules per second. This is a power of 13,750 *watts* (or 13.75 *kilowatts*). The *watt*, named for James Watt, the inventor of the piston steam engine, is a measure of power.

We have just computed the time rate of energy consumption, or the power, the car is consuming when it is traveling at about 30 meters per second (60 miles per hour). To sustain the motion of the car, the engine must produce this much power by burning fuel. This is about equivalent to the power consumed by 137 one-hundred-watt light bulbs.[11]

We might ask, "Where did all that lost energy go?" If you ask that question, then you have indeed learned our all-important lesson about energy—*energy is conserved and cannot be created or destroyed*—it therefore must have gone somewhere else. In the case of our car, the kinetic energy is lost, through friction between mechanical parts, into heating the engine, through the sound energy the car produces, into the energy content of the air that the car moves around as it travels down the highway, and into the energy of heating and compressing and deforming the tires as they spin around. However, most of the wasted energy goes into heat, increasing the speed of the molecules of water (engine coolant), tires, the road, and so on. Since this is chaotic and random molecular motion, it is virtually impossible to recover this energy usefully.

We living organisms are also engines. Our bodies are consuming energy to sustain our metabolism, ergo our lives. Here we measure energy in "food calories," usually designated with the uppercase C, as in the word *Calorie*. A typical (lean) person in the United States eats about 2,000 Calories per day. To convert this into joules we multiply by (approximately) 4,200; hence, the average lean person is consuming about 8,400,000, or 8.4 *million*, joules of food energy per day! In a day there are 24 hours and 60 minutes per hour, and 60 seconds per minute, that is, 86,400 seconds total in a day. Therefore the average person consumes energy, and burns off the equivalent energy, at an average rate of about 8,400,000/86,400 = 97 watts. Therefore each of us, as living, functioning, metabolizing beings is approximately equivalent to a 100-watt light bulb in our metabolic power consumption.

THE LOOMING ENERGY CRISIS

Yet most Americans, in their everyday lives, are consuming far more energy than the mere 100 watts needed to survive. On average, in their homes about 3,000 watts of power (or 3,000 joules of energy per second[12]) is continuously being consumed. This includes electrical lights, heat, refrigerators, air conditioners, televisions, etc. Moreover, Americans consume about 10,000 watts *per person* when we include our automobiles, trucks, airplanes, transportation, factories, energy losses in power transmission, illumination of office buildings, and large aircraft carrier battle groups (to maintain the commercial flow of oil that we burn to create part of this power in the first place). The American per capita power consumption rate is about five times greater than the world average. With better technology, and a change in our behavior, it could be significantly reduced.

For comparison, the Sun produces a power of about 100 watts per square meter at ground level on Earth, averaged over a sunny day. So, a solar collector 300 meters square, about the size of a large roof, with an efficiency of 10 percent, would be required, on average, for every household in our society to obtain from the Sun all the power presently required. Such solar collectors are now less efficient than 10 percent and very expensive, but there are serious efforts under way to make them more cost-effective. If we could live in energy equilibrium, consuming no more energy than is delivered to us from the Sun and producing no toxic waste in the process, our energy problems would be solved. Solar energy, however, may not be practical when a society is consuming as much energy per person per day as ours is.

What about hydroelectric power, and things that depend upon energy storage with gravity, such as tidal ponds? A tidal pond is filled with water during high tide and drains during low tide. It is something like the setup of the Acme Power Company but uses the tides as an energy source. The water is passed through turbine generators to produce electricity, but unfortunately, we find that a tidal pond would have to be enormous to produce sufficient energy for a medium-sized city.

Let's do what is called an "order-of-magnitude estimate." Suppose we take a large region off the coastline of the United States, assumed to be approximately 1,000 kilometers (10^6 meters) in length and construct a "tidal basin" 10 kilometers (10^4 meters) wide, hence with an area of 10,000 square kilometers (or 10^{10} square meters). We'll assume the amount of water that enters the tidal basin in one daily cycle corresponds to a height change of 1 meter, hence a volume of water of 10 billion cubic meters (10^{10} cubic meters). This has a mass of 10,000 billion kilograms (10^{13} kilograms; since 1 gram is the mass of one cubic centimeter of water, and we have 10^{10} cubic meters times 10^6 cubic centimeters per cubic meter, the total is 10^{16} grams $= 10^{13}$ kilograms). The water is hoisted by the tides 1 meter and then released during the daily cycle (actually, the average hoisting distance is 0.5 meters, but we are only doing an order-of-magnitude estimate here, so we round this up to 1 meter for the sake of simplicity). The gravitational potential energy is recaptured by allowing the water to flow out of the tidal basin through the turbine generators, and we'll assume 100 percent efficiency for the conversion to electricity. The gravitational potential energy involved, *mgh*, using the fact that $g = 10$ meters per second-squared, is therefore 100,000 billion joules (10^{14} joules, or 10^5 gigajoules, where the prefix *giga-* means "billion"). Dividing by the number of seconds in the daily cycle, approximately 100,000 (10^5), we see that this can produce an average power output of about 1 gigawatt, that is, a power of a billion watts. This power output can provide sufficient energy for only three hundred thousand people at three thousand watts per capita consumption. The tidal pond energy is essentially free but, as we see, requires a vast area of sea that must be enclosed by the tidal basin. We have also assumed 100 percent efficiency in this estimate, which is overly optimistic. The tidal basin we've described could produce the power needs for a large town, but could not produce the power needs for New York City, for example.[13]

A novel technology of note is the so-called pebble bed nuclear reactor. These reactors, like nature's own Oklo, consume fissionable uranium. The uranium is prefabricated into billiard ball–sized elements encased in glass seals, making them chemically inert (these systems cannot utilize the more chemically reactive plutonium). They produce electrical energy by superheating helium gas, which is then passed through turbine generators to make electricity. There are key advantages to these systems over other nuclear reactors: for one, safety. Helium, unlike water, is chemically inert and doesn't degrade the housing and pipes it passes through. After the fuel is exhausted, the billiard balls are hauled away but must be stored in a nuclear waste facility, and not many people want these in their backyards. Pebble bed nuclear reactors, nonetheless, are one of the least expensive known ways to produce electrical power today. These facilities can be inexpensively prefabricated, and a single unit, of about the size of a barn silo, can produce one hundred megawatts, enough to power thirty thousand people at the consumption rate of three thousand watts per person. However, it should be kept in mind that the consumption of fissionable uranium for energy production on a large scale simply buys time—eventually, on a time scale comparable to oil, it will be consumed as well.

Wind farms have recently become a central focus in the discussion of energy policy. One can construct systems of a very large scale, such as windmills 100 meters (300 feet) tall from the blade tip to the ground, which can produce about 1 megawatt each at a high wind speed of 10 meters/second (hence, one hundred such windmills can produce the same amount of energy as the pebble bed nuclear reactor we just analyzed). These systems have become competitive with fossil fuel energy sources due to the use of modern materials, which make them hardier in fierce storms. They can be located offshore, where higher wind speeds are constant. Europe's experience has suggested that there are few problems with this technology. There are, however, aesthetic issues, and a resistance to peppering the landscape—on- or offshore—with large, noisy, behemoth windmills is growing. Cities such as Copenhagen, however, have become models for the utilization of offshore wind farms and acceptance by the population.

And, finally, what about nuclear fusion? This, you recall, is the energy that powers the stars and leads to the formation of ordinary matter. All atomic nuclei lighter than iron can be produced by fusion—the combining together of two nuclei to make a heavier nucleus, with the release of energy. A typical process is the combining of two deuterium nuclei (deuterium is a nucleus with one proton and one neutron, therefore an isotope of hydrogen) to make helium. Indeed, *all* of our energy sources are essentially derived from fusion, since everything we burn or eat or extract from natural sources was created by the energy output of the Sun, and that energy source is nuclear fusion. Even the fissionable materials used in nuclear reactors were created in the powerful explosions of supernovas, dousing heavy elements in a sea of neutrons, in the terminal stage of a star that had produced its heavy iron core through fusion.

Though the early promises to create virtually infinite energy sources from nuclear fusion have not yet panned out, it is far too early to give up on this potential final solution to the energy needs of humanity. It is a long-term research challenge to solve the problem of harnessing nuclear fusion, perhaps requiring another forty or fifty years to demonstrate feasibility, and it won't be cheap—perhaps costing as much as one annual US defense budget. Large-scale nuclear-fusion research has become an unprecedented international scientific collaboration, and the current keystone project, aimed at demonstrating the scientific and technical feasibility of fusion energy by 2050, is the International Thermonuclear Experimental Reactor (ITER).[14] Many other, smaller innovative efforts are under way worldwide. Let's keep our fingers crossed.

One common feature emerges from our brief analysis: we see that much of the problem lies in the high rate of energy consumption. Therefore it may ultimately be a prudent idea to reduce the consumption of energy in our society, through modern technology and better governmental energy policies. By maintaining gas-guzzling automobiles, poor energy conservation technologies, deficient or altogether absent mass transportation systems, absolute recklessness in our energy behavior, and no sensible energy policies from government, we could easily keep the present rate of energy consumption as it is. Nonetheless, if policy, pollution, and global warming don't change our behavior, the law of conservation of energy eventually will.

There is much more to say about energy; we have hardly scratched the surface. Why does such a thing as "energy" exist at all? What are its deep connections to the symmetries of the laws of physics? Why is there a connection between energy and time? Let us explore further.

chapter 3

EMMY NOETHER

It would be very helpful for the purpose of psychological investigation to know what internal or mental images, what kind of "internal words" mathematicians make use of.

—Albert Einstein

The universe displays a great deal of symmetry. For example, there is no center about which everything turns in the universe. Any point in space is just as good a "center" as any other. The more profound statement is that the laws of physics themselves do not depend upon where you are in empty space. Furthermore, they do not depend upon one's orientation in empty space. For instance, rotating an experiment into a different orientation in space won't affect its outcome. Nor, as we have seen, do the laws depend upon what time it is. Time, like space, has no special or preferred point. One might think that a possible exception is the initial instant of the big bang. But even the big bang event of cosmology, as far as we know, was governed by the same laws of physics that govern the formation of a raindrop over a cornfield in Kansas. The laws of physics govern the structure of space and time, even under extreme conditions. The initial instant of the big bang is the way in which time and space behaved, according to these laws, when the universe was composed of matter at a virtually infinite density.[1]

There are likely to be hidden symmetries—symmetries we haven't yet detected. There may even be many other universes, perhaps connected to ours through such phenomena as *space-time tunnels*. Time may have existed in another universe where the notion of "before or after ours began" loses meaning. There may be many other, unseen dimensions of space and time, so small that our particle accelerator microscopes cannot yet detect them. These dimensions would possess similar symmetries to the ones we see, with no particular direction in space that is special or preferred. Or the new dimensions may involve peculiar new abstract mathematical numbers with strange properties (like so-called anti-commutation, where $3 \times 4 = -4 \times 3$). These would lead to symmetry principles that govern the properties of matter called *supersymmetries*, and they would predict new forces and new elementary particles.

Thus there is nothing special about "up" or "down" or "sideways" or "forward" or "backward" in the laws of physics. The universe is perfectly democratic; all directions, places, and times are created equal, and much more symmetry exists beyond that, reflected within the world of the elementary particles and forces, the basic component parts of nature.

The constancy of the basic parameters, the "fundamental constants" of physics, over vast distances and times has been established in astronomical and geological observations to a precision of approximately 1/1,000,000,000, as we have seen in the ancient natural nuclear reactor at Oklo and elsewhere. It is important to recognize that the basic laws of physics are also the same over extremely small distance and time scales, the fleeting time scales of the processes involving the elementary particles.

Any constancy, or invariance, is considered by physicists to be a symmetry of nature. The symmetry can be the equality of the laws as we move from place to place in the universe, in space, time, or orientation. It can be the constancy of the properties of a system as we do something to try to change it. That the laws will be the same tomorrow as they were yesterday is a symmetry as we drift forward in time within the universe. That the laws will be the same elsewhere as they are here is a symmetry as we move about space within the universe.

MATHEMATICS VERSUS PHYSICS

Emmy Noether is regarded as the greatest of all woman mathematicians, and one of the greatest mathematicians—man or woman—in history. She developed entirely new branches of *abstract algebra*, which pushed the envelope of the world of mathematical systems, refining the nature of what mathematics is. She created a remarkable and celebrated algebraic system known as Noetherian Rings. Yet her most profound contribution was to theoretical physics, and ultimately to our understanding at the deepest levels of how the universe works. It is probably true, even today, that although most mathematicians may have heard of Noether's theorem in physics, they are not fully aware of its significance to theoretical physics. It is certainly true that most physicists have no real appreciation or knowledge of Noetherian Rings in mathematics. The worlds inhabited by theoretical physicists and mathematicians are often quite separate and independent. It is during the rare moments when the two worlds converge that the bugles blow, the drums roll, and science moves forward!

If we ask a class of high school students to name the most famous professional basketball players, we can very quickly generate a fairly long list. Still, we often hear the name Michael Jordan first (especially in Chicagoland). However, when the same students are asked to name the greatest mathematicians who ever lived, the list's length grows at the rate of a tennis ball sinking in a jar of molasses. Nevertheless, we usually hear the name Albert Einstein first. We don't hesitate to give credit for that answer, but do we reflect on where the full impact of Einstein's revolution mainly lies? Indeed, it lies not in mathematics but in theoretical physics. Theoretical physics borrows from mathematics (or, if there's none to borrow, they invent new mathematics) in order to create a mathematical roadmap of things that can happen in the real world, in nature. It strives to explain all of the many different phenomena observed in the universe, perhaps ultimately seeking one elegant and economical logical system. However, physicists usually settle for lesser triumphs, in which many physical systems with common and comprehensible behaviors are successfully described. This description is always created in the abstract language of mathematics.

Nature has indeed revealed deep mathematical underpinnings and interrelationships among various phenomena. For example, we learned in the mid-nineteenth century that magnetism is related to the electric force—through motion—that they are, in fact, one and the same when unified by the symmetry of the laws of nature under different states of motion. We call this phenomenon *electromagnetism*, and all electromagnetic phenomena are neatly summarized in one reasonably simple and beautiful theory, containing a remarkable amount of symmetry. This unified description of electromagnetism is usually referred to as "Maxwell's equations," named after British physicist James Clerk Maxwell (1831–79). But there are many alternative and equivalent mathematical ways to formulate the description of the related phenomena, just as there are many ways to paint a majestic mountain landscape. The main point is that, as far as we can tell, all natural phenomena ever observed are bridled by the deep logic of mathematics. Nature seems to speak the language of mathematics.

Although nature suggests a pathway to a mathematical description of everything, it has thus far eluded a final or complete grand mathematical synthesis. There has been enormous progress in the area of *superstring theory*, in the description of all forces and particles in one mathematical theory including gravity as its centerpiece. However, many loose ends still exist. The theoretically anticipated pattern of elementary particles known as supersymmetry, stemming from superstring theory, may someday (soon?) begin to emerge in experiment. Or else, perhaps we will receive a stern memo from the supreme being(s) that mere human minds cannot anticipate the totality of nature's diversity in the absence of experimental evidence. In any case, there is much yet to be done, in both theoretical and experimental physics, and many Nobel Prizes for young or future generations of physicists remain to be awarded.

The subject of mathematics has its own identity, on the other hand, and, in contrast to physicists, mathematicians attempt to create a roadmap of all possible *logical systems* that could consistently exist, whether they ultimately have anything to do with nature or not. Yet nature provides the basis of abstraction that gives birth to mathematics. The forms seen in nature—triangles, circles, polygons, polyhedrons—were abstracted by the Greeks into the construction of the first complete mathematical system, Euclidean geometry. Mathematics is therefore inspired by nature. But it does not have to conduct experimental observations to proceed. The worlds of mathematics and theoretical physics are therefore distinct—they have different "mission statements." Whereas theoretical physics maps the properties of the nature we experience, mathematics builds a map of all possible "natures" that logic permits to exist. So it is that the worlds of mathematics and theoretical physics coexist and flourish, sometimes in Golden Ages and sometimes not. Like a married couple in an old apartment building in Manhattan, sometimes we hear them fight, and other times we hear lovemaking through the old plaster walls. Mostly we hear silence and peaceful coexistence.

An asymmetry of purpose and direction thus exists in these two worlds of the intellect, mathematics and theoretical physics. Emmy Noether's greatest contribution, Noether's theorem, is the powerful connection that leaps from one to the other, like a tunnel connecting two different universes. It is the doorway between symmetry and the dynamic behavior of physical systems.

THE LIFE AND TIMES OF EMMY NOETHER

Emmy Noether practiced in the time of profound new revelations about the very structure and shape of mathematics. It was the early twentieth century, and she participated in this era of radical revisionism and grand synthesis for both the fields of theoretical physics and mathematics. Both fields of mathematics and physics were making maps of newly discovered territories and revising the centuries-old maps of the classical era. The style and judgment brought to bear were related, yet quite different, in these two disciplines.

Max Noether, Emmy's father, was one of the finest mathematicians of the nineteenth century. Germany was then the intellectual center of mathematics, as well as all the physical sciences, engineering, medicine, and biology, and it was the center of rapid technological advancement. This was a period of staggering change in the German society, politically, territorially, culturally, economically, and socially. It was the era of the German Empire and its powerful chancellor Otto von Bismarck-Schönhausen (1815–98), who led the unification of hundreds of assorted tiny principalities and states to forge a single, powerful new nation.

At this time there was wealth, freedom, and tolerance in Germany, with a steady improvement in the plight of the common people, due mainly to the Industrial Revolution. An optimism prevailed that fostered the belief that human beings might ultimately achieve "utopia." Crown Prince Friederich II, son of the reigning kaiser Wilhelm I and the father of the future kaiser Wilhelm II, planned to introduce many socialist reforms to better the lives of the poorer classes, especially the miners, in the emergent nation and under his own rule. This may have happened, and the subsequent twentieth century may have enjoyed an entirely different outcome, had Friederich not succumbed to an untimely death by throat cancer before he was able to become kaiser. A mild and treatable cancerous lesion on the crown prince's throat became fatally exacerbated when he chose to follow the "bed rest" order of the leading British laryngologist of the time (highly recommended by Queen Victoria herself), rather than the urgent pleas of his skilled German surgeons to cut the lesion out. The succession passed onto his son, Wilhelm II, in 1888, and the seeds were planted for an eventual turn down the path toward the Holocaust.[2]

The Noethers were Jewish, an ethnic minority in Germany that had long suffered persecution throughout northern Europe. Max Noether was born in Mannheim in 1844 into a family that owned and operated a successful wholesale hardware business. He was stricken with polio at the age of fourteen, an illness that handicapped him for the rest of his life. He managed to complete his studies at *Gymnasium*, nominally equivalent to a traditional US high school, through study at home. Like many of the great mathematicians, he began studying advanced mathematics independently and was largely self-taught. This approach provided him the time to focus on the tricky points and to travel at a pace that worked best for him as an individual. Max Noether later entered the University of Heidelberg in 1865, receiving the equivalent of the PhD degree, within the short period of three years.

The university of late nineteenth- to early twentieth-century Germany was a place of many contrasts and contradictions. It was a profoundly influential community, particularly in the sciences and mathematics, where it enjoyed the reputation of being the best in the world. It was a place of the highest academic standards of the age, the birthplace of quantum mechanics and Einstein's general theory of relativity as well as most of modern mathematics, such as abstract algebra, topology, differential geometry, and others. Here ethnic minorities found a tolerant, open, and receptive community, a place to flourish that offered a respite from an outside society of staunch national conservatism. This was a quiet, meditative environment, a community of scholars with a common deep and abiding love of their abstract pursuits. German universities, however, were also places in which mainstream "educable" young boys were sent from their homes to become men, usually in conjunction with some military service.

Max Noether was a quiet scholar, leading the scholar's ascetic life, disjointed from mainstream German society. He served on the faculty at the University of Heidelberg for several years. He then joined the faculty at the University of Erlangen and served as a chaired full professor there from 1888 to 1919. He is recognized as one of the founders of nineteenth-century *algebraic geometry*, following in the footsteps of another great mathematician, Bernard Riemann, one of the fathers of *non-Euclidean geometry*, later to become the basis of Einstein's general relativity.

In 1880 Max Noether married Ida Amalia Kaufmann. Their daughter, Amalia, or "Emmy," was born March 23 in 1882, named for her mother. Emmy, who had three younger brothers, attended grammar school in Erlangen in the 1890s, studying languages, mathematics, and the piano. Her early ambition was to become a language teacher.

Instead of pursuing a teaching career, however, Emmy abruptly changed direction. She decided to concentrate her advanced study in the chosen career of her father, the field of mathematics. This was essentially unheard of for the women of her day. Women were generally permitted only to study, unofficially, at German universities, requiring each professor to give permission to attend his course. Despite these hurdles, Emmy Noether completed her course work and passed the official matriculation examination in 1903, which is roughly akin to getting a bachelor's degree.

Emmy Noether then went on to the University of Göttingen for her graduate-level work. Here she attended lectures by the great mathematicians of the age: David Hilbert, Felix Klein, and Hermann Minkowski. She completed her doctorate there in 1907 but then returned home to Erlangen to help her by then aging and frail father. At this time she also began her own research career in mathematics. Her reputation as a brilliant mathematician began to spread immediately, and during this period, she received numerous honors and accolades.

David Hilbert was the leading figure in mathematics in the early twentieth century. Indeed, at the beginning of the century, certain logical contradictions had been found in *set theory*, which is an abstract system for discussing the logical structure of all of mathematics. Up to that time, set theory was viewed as the fundamental natural foundation of mathematics. Hilbert proposed a program to straighten out these logical structural problems and to "clean up" mathematics.

Hilbert's program for mathematics was presented in a famous speech, titled "The Problems in Mathematics," that he delivered in 1904 to the Second International Congress of Mathematicians in Paris. Here he challenged the worldwide community of mathematicians to focus upon and solve twenty-three exemplary problems that he considered to be the most fundamental or illustrative questions—things that would best illuminate the internal structure of mathematics itself. These included the *Riemann hypothesis*, the *continuum hypothesis*, and *Goldbach's conjecture*, among others. Many of these celebrated problems are unsolved to this day, and their proof remains the cause célèbre of the global mathematical community. Others were solved during the latter part of the twentieth century—and some are probably being solved at this very moment. Hilbert felt that, finally, a consistent mathematical edifice could be built and that mathematics could be understood to be a regular and complete logical system. Thus, like a knight on an infinite chessboard, with sufficiently many logical moves one could always get from one to any other square on the chessboard. Hilbert believed no hidden surprises remained in mathematics.

In 1915 Hilbert and Klein invited Emmy Noether to return to the University of Göttingen to conduct her research and teach. Noether worked at Göttingen but in an ill-defined, unsalaried, and subordinate role, while Hilbert fought strenuous battles with the university authorities to allow a woman to become a member of the faculty. A majority of faculty members argued against this: "How can it be allowed that a woman should become a *Privatdozent* [roughly an academic rank equivalent to that of assistant professor]? Having become a *Privatdozent*, she can become a Professor and a member of the University Senate. . . . What will our soldiers think when they return to the university and find that they are required to learn at the feet of a woman?"

To this argument Hilbert replied, "Gentlemen, I do not see that the sex of the candidate is an argument against her admission as a *Privatdozent*. After all, the senate is not a bathhouse."[3]

It should be noted that such chauvinism toward female academicians was not unique to Germany in the history of mathematics and science. At issue was whether Noether could receive the *Habilitation*, a qualification required of scholars before being admitted to lecture at a university—but it was not permitted to grant this to a woman. But in 1919, by the efforts of Hilbert and the remarkable talents of Noether, permission for the *Habilitation* was finally granted. Throughout this time, Emmy Noether gave public lectures that were advertised in the name of Professor Hilbert:

MATHEMATICAL PHYSICS SEMINAR:
PROFESSOR HILBERT, WITH THE ASSISTANCE OF
DR. E. NOETHER,
MONDAYS FROM 4–6, NO TUITION.[4]

Emmy Noether's first work in 1915, shortly after she arrived in Göttingen, was her profound work in theoretical physics, proving Noether's theorem. In short, the theorem states that for every continuous symmetry in the laws of physics, there is a corresponding conservation law. As we have seen in one example of the theorem at work, the symmetry, or invariance, of the laws of physics with time leads to the *conservation law of energy*. It also works in reverse: the conservation of energy implies that physics is not changing in time. However, her theorem goes well beyond energy conservation, highlighting in a profound and fundamental way that symmetry is the underlying and most important theme of nature. We will expand upon the meaning and multitude of implications of this elegant and simple result throughout this book. All conservation laws deeply reflect fundamental symmetries of the laws of nature. Noether's theorem succinctly codified and unified many ideas that had been around for a long time and placed them firmly upon the pillars of symmetry.[5]

Symmetry was a totally modern and revolutionary way to think about the laws of nature. Noether's theorem intimately intertwines dynamics together with symmetry. It ultimately explains the forces and dynamics of nature that arise as a consequence of deep, underlying symmetries. Noether's theorem is certainly one of the most important mathematical theorems ever proved in guiding the development of modern physics, possibly on a par with the Pythagorean theorem. It doesn't lie in the province of mathematics alone but rather is a profound statement about the entire physical world.

Noether's work was immediately recognized to be of fundamental importance. Albert Einstein praised her contribution, describing it as "penetrating mathematical thinking" in a letter written to Hilbert to help boost the career of this gifted young female mathematician. The theorem may have played a role in stimulating David Hilbert's own foray into theoretical physics, when he proposed a formulation of gravity that is virtually equivalent to Einstein's general theory of relativity and arguably contemporaneous.

Noether continued to develop a stunning career in mathematics, which earned her increasing worldwide accolades, ultimately ranking as one of the greatest mathematicians in history. At Göttingen, after 1919, Noether concentrated on the subject, within pure mathematics, of abstract algebra. She made major contributions to this subject, helping to forge *ring theory* into a major branch of mathematics. Ring theory deals with abstractions of numbers, as well as the functions and operations that can be performed upon them. It attempts to distill the structure of algebra down to sets of rules in which the mathematics is determined, irrespective of the details of these rules. Noether's 1921 work "Idealtheorie in Ringbereichen" was of foundational importance in the development of modern algebra.[6] In this paper she gave an illuminating analysis of the fundamental structure of certain algebraic objects, generalizing an important theorem proved earlier by world chess champion Emmanuel Lasker, who was also a student of Hilbert.

Throughout the 1920s, Noether continued with her fundamental work on abstract algebra. In 1924, noted mathematician B. L. van der Waerden came to Göttingen and spent a year studying with her. After returning to Amsterdam he wrote an influential book, *Moderne Algebra*, which appeared in two volumes, the second of which consists almost exclusively of Emmy Noether's work. From 1927 on, Noether continued her collaborations with other noted mathematicians throughout Europe and became an editor of *Mathematische Annalen*, the most prestigious mathematical journal of the era. Much of Emmy Noether's work ultimately appears in the famous papers written by her colleagues and students rather than under her own name. Throughout her career she was a patient and nurturing mentor, known to be generous to her students with many of her new and innovative ideas. It is said that many of her students ultimately received credit for her ideas, which she communicated freely and without academic "liens," so she could help them develop their own careers.

Nonetheless, her fame took her far across the academic globe throughout this period. In 1928 and 1929 she was a visiting professor at the University of Moscow and received a prestigious invitation to address the International Mathematical Congress at Bologna in 1928. In 1930 she taught at the University of Frankfurt, and in 1932 she was again asked to give a lecture at the celebrated International Mathematical Congress in Zurich, where she was awarded the prestigious Alfred Ackermann-Teubner Memorial Prize for the Advancement of Mathematical Knowledge.

In the meantime, David Hilbert's serenely confident view of the completeness and regularity of the house of mathematics was dashed to bits by a radical theorem proved in 1931 by the young Kurt Gödel. Hilbert believed that the whole of mathematics was a completely self-consistent logical system. That is, no theorem could ever lead to an inconsistency with any other. If you prove that I could never walk from the island of

Oahu, Hawaii, to Los Angeles, California, without getting wet, then I cannot prove that there is some hidden land bridge or tunnel between that island and the West Coast over which I can walk and remain dry.

Kurt Gödel revealed that any mathematical system is always incomplete through his famous incompleteness theorem. That is, there are always questions that can be posed in any mathematical structure that cannot be proved to be true or false. This, at some point, must also carry implications for the enterprise of theoretical physics in any quest to finally reduce all of nature into a basic set of defining equations. It would naively seem to imply that there is always some experiment that can be performed with a definite outcome that cannot be predicted by the mathematics of theoretical physics.

Mathematics is therefore not a simple roadmap or a chessboard, with straightforward rules that allow the knight to move ultimately between any two squares. Gödel essentially showed that there is a square somewhere on the chessboard of any mathematical system that the knight can never reach! Mathematics itself defies a complete mathematical analysis. Rather, its structure is more chaotic and unmappable, and any two seemingly neighboring points may in fact be completely disconnected from each other. No logical proof exists of all the theorems one can pose in a mathematical system.[7] The knight cannot visit all of the squares of the hypothetical mathematical chessboard!

Unfortunately, much more than the house of mathematics was thrown into chaos in the early 1930s. So, too, was the peaceful, seemingly utopian life of German academia. In 1933 the rise of the dark storm clouds of Nazism in Germany led to Emmy Noether's dismissal, together with that of all other ethnic minorities, from the University of Göttingen. The Prussian ministry for science had published a list of professors who had Jewish ancestry, and Emmy Noether was on the list. Within a few days they were all fired, gutting the renowned mathematics and physics departments of the greatest universities of Germany. For a while, Emmy conducted an underground mathematics class in her own apartment for her students, but often the topic of discussion turned to the current events. Hermann Weyl wrote about her during this period: "A stormy time of struggle like this one we spent in Göttingen in the summer of 1933 draws people closely together; thus I have a vivid recollection of these months. Emmy Noether . . . her courage, her frankness, her unconcern about her own fate, her conciliatory spirit . . . was, in the midst of all the hatred and meanness, despair and sorrow surrounding us, a moral solace."[8]

Noether was invited to visit the United States for the academic year of 1934 and accepted a visiting professorship at Bryn Mawr College (see fig. 3). She also often lectured at Princeton during this period. During the summer of 1934, she returned again to Göttingen to close up her apartment and ship her belongings to Bryn Mawr. It was then she bade her final farewell to her family and friends. One cannot help but wonder about the fates of her family and friends, ones who remained behind, many in denial about what horrors might lie ahead.

For Emmy, what appeared to be a happy new life at Bryn Mawr was to be cut short. She was diagnosed with a large ovarian tumor in 1935 and underwent surgery on April 10. Four days after the surgery she suddenly lapsed into a coma, and her temperature ran up to 109 degrees. She died on April 14, 1935, at the age of 57, the specific cause having been diagnosed as a stroke.

Emmy Noether was quoted as saying that the last 1½ years of her life had been the very happiest. She had acquired new friends and was welcomed and appreciated at Bryn Mawr, and at Princeton, as she had never experienced in her own country. Perhaps the gloom of Europe in the previous decade had worn her down. She was spared the knowledge of the fates of close friends and relatives in the Holocaust, and the specter of the great academia of nineteenth-century Germany, her father's peaceful world, burned to the ground.[9]

In a tribute, Albert Einstein wrote in the *New York Times* on May 4, 1935:

> There is, fortunately, a minority composed of those who recognize early in their lives that the most beautiful and satisfying experiences open to humankind are not derived from the outside, but are bound up with the development of the individual's own feeling, thinking and acting. The genuine artists, investigators and thinkers have always been persons of this kind. However inconspicuously the life of these individuals runs its course, nonetheless the fruits of their endeavors are the most valuable contributions which one generation can make to its successors. . . .
>
> . . . In the realm of algebra, in which the most gifted mathematicians have been busy for centuries, [Noether] discovered methods which have proved of enormous importance. . . . Pure mathematics is, in its way, the poetry of logical ideas. . . . In this effort toward logical beauty, spiritual formulas are discovered necessary for deeper penetration into the laws of nature. . . . The efforts of most human beings are consumed in the struggle for their daily bread, but most of those who are, either through fortune or some special gift, relieved of this struggle are largely absorbed in further improving their worldly lot.[10]

Figure 3. Photos of Emmy Noether, circa 1932–33, when she was coming to the United States as visiting professor of mathematics at Bryn Mawr College, a time that Noether described as the happiest in her life. (Photos courtesy of the Bryn Mawr College archives.)

In 1993 the city of Erlangen dedicated a newly constructed school to her, the Emmy Noether Gymnasium. In addition, in 1993 the collected works of Emmy Noether were published. Her ashes were buried under a brick walk in the library cloisters in the school dedicated to her on the hundredth anniversary of her birth, on the occasion of a symposium held by the Association of Women in Mathematics.

SYMMETRY AND PHYSICS

The remarkable connection between symmetry and physics is a modern concept that developed largely in the twentieth century. Physicists in earlier times largely viewed the physical world as composed of "gears and pulleys." Even James Clerk Maxwell, who played a major role in formulating the theory of electromagnetism, thought of the world as a purely kinetic system. Physicists prior to the twentieth century generally didn't think in terms of fundamental underlying symmetry principles. They tended to view symmetry as more of a sideshow, a toy, arising in an occasional situation involving a symmetrical configuration that could help simplify a specific physics problem but that played no profound role in the deeper dynamic fabric of the physical world.

It was Albert Einstein who brought in a new kind of thinking with his development of the theory of special relativity. Einstein thought at a deep level about the symmetries of space and time. He discovered special relativity hiding within the equations of Maxwell's theory of electrodynamics. This was possible only through the new perspective that Einstein brought to bear. As we will see, relativity actually began with Galileo and is all about the symmetry of space and time. Yet Einstein's perspective was modern: he sought a kind of underlying naturalness to extract the true laws of physics and discovered far deeper principles of symmetry than had been seen before. Noether's theorem was born out of this new perspective.

As we will see, the mere existence of certain symmetries requires the existence of the forces that we observe in nature. We now know that all forces in nature come from these deeper kinds of symmetries, called gauge symmetries. The idea of underlying symmetries and Noether's theorem have led ultimately to the discovery of the unifying principle governing all the known forces in nature. Understanding the principles of local gauge symmetry has allowed us to leap conceptually to distance scales 100,000 trillion times (10^{-17}) smaller than can be seen with our most powerful particle accelerators, the most powerful microscopes humans have ever built. At such short distances quantum gravity is active and defies our normal notions of space and time. Still, we can carry along, on this venture, our concepts and principles of symmetry. There we must use our symmetry principles to imagine the complete unification of all forces into something like superstrings, one of the most symmetry-laden systems ever conceived of by the human mind.

Physicists now revere these abstract yet fundamental symmetries of nature, and we have come to see them as real and to intimately appreciate their subtle consequences. To succumb to the pipe dream of a perpetual-motion machine would require us to give up the law of energy conservation. We would be forced to give up the notion that the flow of time is a symmetry in which there are no changes in the laws of physics. Indeed, as we will see, symmetry controls nature in a most profound way. This was the ultimate lesson of the twentieth century for the children of the Titans.

chapter 4

SYMMETRY, SPACE, AND TIME

> *Symmetry, as wide or as narrow as you define its meaning, is one idea by which man through the ages has tried to comprehend and create order, beauty and perfection.*
>
> —Hermann Weyl, *Symmetry* (1952)

The space and time of the universe that we humans inhabit contain symmetries. These are almost obvious yet subtle, even mysterious. Space and time form the stage upon which the *dynamics*—that is, the motion and interactions of the physical systems, atoms, atomic nuclei, protozoa, and people—are played out. The symmetries of space and time control the dynamics of the physical interactions of matter.

We humans live in a three-dimensional universe of space, with one additional dimension of time. We can evidently freely travel *continuously* through space in any direction. All directions in space appear to us to be equivalent. Unlike a chessboard, where a chess piece must take a discrete step to hop to the next square, there is apparently no smallest nonzero step we must take (that we can detect) in space to move around. We see no evidence, for example, that the space in our universe is a lattice, that is, a regular periodic array of points. Time, likewise, flows continuously and not in discrete steps like the tick-tock of a clock. Our space and time appears to be a *continuum*.

How do we know what the fundamental symmetries of space and time are? How can we test that they are indeed symmetries? How do we know, beyond what our eyes tell us, whether any apparent symmetries hold at all distance scales? How do we know that space and time form a continuum? Does the world actually change and become a discrete chessboardlike structure, perhaps like a crystal lattice, at subatomic distance scales, or is it continuous at all scales of distance and time?

GEDANKENLAB

We can imagine having performed a collection of hypothetical experiments that addressed these questions. Physicists often use the German term *Gedankenexperiment*, which means literally a "thought experiment," for such hypothetical exercises. Let us imagine that we know of a very sophisticated laboratory that we'll call "Gedankenlab" (see fig. 6). Gedankenlab was sent out into a vast region of empty space, and it has an enormous amount of time allocated for these experiments. The laboratory had an unlimited mission to conduct various experiments at different locations throughout the space and time that fills our universe.

Gedankenlab blasted off to measure the *fundamental constants*, or *parameters*, that enter all of the equations of physics—the equations that allow us to predict how something will behave under a given set of circumstances. The basic parameters were carefully and precisely measured throughout the universe.[1]

One of the many things Gedankenlab measured throughout our universe is the *speed of light*. Hence, Gedankenlab conducted various measurements of the speed of light in different places throughout space, comparing the results as it moved about. Gedankenlab compared the results of its measurements between different points in space, separated by vast distances. And, with its powerful microscopes and accelerators, it compared the laws of physics between points in space separated by subsubatomic distances, in fact subquark distances. It carefully recorded the time(s) at which these experiments were done.

It conducted the experiments at many different instants and ages throughout the life span of the universe, at the very beginning of the universe (during most of the evolution of the universe), throughout the his-

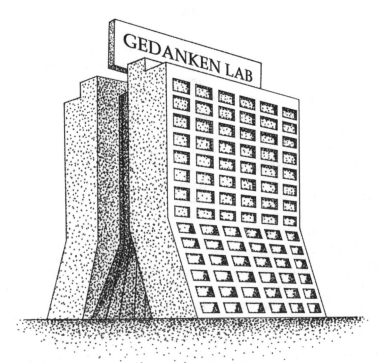

Figure 4. Gedankenlab. (Illustration by Shea Ferrell.)

tory of the universe, and also with great precision over very short time differences. The laboratory had thruster rockets so it could rotate its orientation relative to the rest of the universe. It carried out experiments, looking for small deviations in the speed of light, in various orientations of the laboratory in space. It tried to find out if there was an "up," "down," "sideways," "backward," or "forward" directional dependence in the observed values of the physical parameters that make up the laws of nature. Was the speed of light the same for light moving "up" as opposed to moving "down"? These are the questions that Gedankenlab tried to answer.

These measurements could, in principle, be done at very short distances by observing the behavior of atoms, nuclei, or the properties of quarks and leptons as they are oriented or moved differently in space. For example, one could measure if an electron moves the same way in a magnetic field, depending upon how the magnetic field is aligned in any given direction in space. Such behavior of the electron depends intrinsically on the speed of light. It indirectly determines if light itself moves at the same speed in space, irrespective of direction in space.

Restated in the fancy language that scientists use among themselves, "Is space *isotropic*?" That is, is space the same in all directions? Or are there "preferred," or special, directions in space? If the speed of light is different for light traveling in a particular direction, let's say toward the north-pole star, Polaris, we would be forced to conclude that space is *non-isotropic*!

The results of the measurements of the speed of light carried out at Gedankenlab were finally compiled and, at a big interplanetary scientific conference, the answers were released to the scientific community. The discovery was a resounding *no!* Gedankenlab found that the speed of light was the same in all directions; hence, *space indeed appeared to be isotropic*. The lab found that this is true for both short distances and large distances. In addition, the lab discovered that the speed of light didn't change with time, and it was the same for all states of motion of the lab. These are symmetries of light, but in a larger sense, they are also the fundamental symmetries of space and time.

Eventually, all the results of all the Gedankenlab experiments were published. The stunning results showed that, to a very high degree of scientific precision, none of the laws of physics depended on where the lab is in space (translations in space) or what time it is (translations in time) or how the lab is oriented (rotations in space). Furthermore, the results for the measurements made inside of Gedankenlab did not depend upon the uniform state of motion of the laboratory—that is, one cannot determine if the laboratory is moving through space or at rest. Evidently, any state of motion, orientation, position, or time is equivalent to any other for the experimental results of Gedankenlab.[2]

Let's examine these fundamental symmetries of space and time in greater detail.

SPATIAL TRANSLATIONS

The ordinary space (of our universe) possesses a continuous translational symmetry—the laws of physics are the same everywhere in space. Space is *not* a crystal lattice or a chessboard with discrete steps for the translations; that is, there is no smallest step for a translation in the space we live, down to the smallest distance we can discern, 1/10,000,000,

000,000,000,000 (or 10^{-19}) meters. By using indirect methods we can infer that space is translationally invariant down to even shorter distances, as small as 1/1,000,000,000,000,000,000,000,000 (or 10^{-24}) meters. Whether this symmetry holds at shorter distances we do not know for certain. Nonetheless, through the application of theoretical ideas and Noether's theorem, there is compelling evidence that it does.

Scientists call space a *continuum*. The idea really comes out of pure mathematics, from the number line composed of *real numbers*. The real numbers include the *rational numbers*, that is, those numbers that can be written as ratios of two integers. In addition, there are *irrational numbers*, such as π and $\sqrt{2}$, which "fill in between" the rationals. There is no definable nearest neighbor for a real number; that is, given the number 3, there is no number that is the closest number to 3. The number line of integers (ordinary counting numbers, 1, 2, 3, etc.) is, on the other hand, *not* a continuum, because there is a unit step between two nearest neighbor integers, such as 6 and 7 (and 3 does have two nearest neighbors among the integers, i.e., 2 and 4).

In ordinary space there is no smallest step through which we can translate a quark or an electron, an atom or a planet in space, down to the tiny distance scales we can see. We thus hypothesize that space has no smallest distance scale. A translation in the continuum of space cannot be thought of as an integer number of discrete smallest steps, because there is no smallest step. In a continuum, the absence of the smallest step implies an infinite number of possible translational symmetry operations. Gedankenlab, on the other hand, had discovered that our universe has three-dimensional continuous translational symmetry. We emphasize that this is based upon observation, and if future experiments with more powerful accelerators, examining space at still shorter distances, reveal an underlying crystal-like structure, then so be it. For now the hypothesis of a continuum, with continuous translational symmetry, appears to be a valid one.

Consider a classroom pointer. Usually this is a wooden stick of a fixed length, about one meter (or a bit longer than one yard). We can translate the pointer freely in space by merely waving it around. Do its physical properties change as we perform this translation? Clearly they do not. The physical material, the atoms, the arrangement of atoms into molecules, into the fibrous material that is wood, and so on, do not vary in any obvious way when we translate the pointer through space. Neither do they change if we point it at a poster of Christina Aguilera, or at the door. Neither the color, nor the length, nor the mass of the pointer changes as we translate it in space. This is a symmetry of the pointer under translations, but it is a larger symmetry of the laws of physics—it demonstrates the statement that the laws of physics themselves are symmetrical under the continuous three-dimensional translations in space. The atoms in the wood do not change in any way when we translate the pointer because the laws that govern them are the same here as they are over there.

Any *mathematical equation* we write describing the quarks, leptons, atoms, molecules, stresses, bulk moduli, electrical resistance, and so on—or just an equation for the length of our pointer—must *itself have symmetry* and be invariant under translation in space; it must be the same equation that applies no matter where we are in empty space. This is a remarkable insight! But what does it mean for an equation itself to have symmetry and to be invariant?

Consider the simplest possible example of an equation having a symmetry. We want to describe the length of our classroom pointer, which we call L. Suppose we have a tape measure that is stretched out alongside the pointer. We can place the pointer anywhere alongside the tape measure to measure its length. All we need to do is measure the position of the tip of the pointer, which is some marking on the tape measure, x_{tip}, and we may find that $x_{tip} = 79$ inches. We also need to measure simultaneously the position of the end of the handle of the pointer, x_{handle}, and we get the answer $x_{handle} = 49$ inches. Therefore we know that the length of the pointer is $79 - 49 = 30$ inches. More generally, the mathematical *formula* for the length of the pointer is just $L = x_{tip} - x_{handle}$.[3]

Now, a friendly high school physics student who lives on our block, Sherman, comes along and fiddles with the tape measure, pushing the yellow button and watching it zip closed several times. He then repeats the measurement of the length by again extending the tape and placing it on the table alongside the pointer. This time, Sherman finds, simultaneously, that $x_{tip} = 54$ inches and $x_{handle} = 24$ inches. It would appear that the measurements of the location of the tip and handle have now changed. This is a result of a *transformation*, or *operation*, that has been performed on the tape measure and pointer system. The pointer has been *translated in space relative to the tape measure*. However, there is a *symmetry*—the symmetry of *translational invariance*. The length of the pointer has not changed; it is still $54 - 24 = 30$ inches.

So, we therefore see that our mathematical formula $L = x_{tip} - x_{handle}$

itself has a symmetry. We can perform an *operation*, which is to *transform* the values of x_{handle} and x_{tip} by a translation in space. To do this we simply replace the values of these quantities by new values (indicated with primes): $x'_{\text{tip}} = x_{\text{tip}} + D$, and $x'_{\text{handle}} = x_{\text{handle}} + D$. The quantity D is the amount by which we have shifted, or translated, the pointer in space relative to the tape measure. But this doesn't affect the result of the formula, which is the length of the pointer: $L = x'_{\text{tip}} - x'_{\text{handle}} = x_{\text{tip}} + D - (x_{\text{handle}} + D) = x_{\text{tip}} - x_{\text{handle}}$. The implication of this exercise is that the final result for the length of the pointer *does not depend* upon the amount of translation, D. D cancels out of the formula, no matter what value it takes. The formula, we say, is *invariant* under the operation of translating the pointer in space. We say that our formula "displays the translational symmetry." Symmetry is present because the equation does not refer to any special point in space, because space has no special points. This has to be true since the equation itself reflects the fact that *the laws of physics are invariant under translations in space.*

TIME TRANSLATIONS

We can consider time to be like space, and we can imagine translating a physical system *in time*. We can study, for example, the properties of a top quark, the heaviest elementary particle ever detected, at the Fermi National Accelerator Laboratory at 9 AM, or we can consider the properties of the top quark at 3 PM. Do the intrinsic properties of a top quark, its mass, its electric charge, and so on depend upon what time it was produced? After performing the experiment, Gedankenlab has hypothetically told us *no!* The properties of the top quark are simply reflecting the laws of physics; hence, we have discovered that *the laws of physics are invariant under translations in time.*

That is to say, the result of any experiment we do tomorrow, or ten seconds ago, or five years in the future, and so forth, will be the same. The laws of physics, and thus all correct equations in physics, are invariant under translations in both space and time. This is an experimental fact, as far as we can discern.

We don't have to take the word of the Gedankenlab scientists alone. Indeed, the global constancy of the basic parameters of physics over vast distances and times has been established in astronomical and geological observations, such as through the production of samarium at the Oklo natural nuclear reactor, to better than approximately one part in ten million precision over the entire lifetime of the universe—some thirteen billion years! Furthermore, Oklo is not alone. There are many other indicators of the stability of the laws of physics through global time. Astronomers can peer through telescopes at faraway stars and galaxies and see that the same physical processes are at work in those distant bodies of long ago, as we find occurring here in our laboratories on Earth today. The abundance of certain elements in meteorites tells us that other very sensitive processes are the same today as they were billions of years ago. In the 1970s, the Viking lander, sent to Mars by NASA, allowed a precise measurement of the force of gravity.[4] It was determined that gravity, too, does not change through time. Taken together, all of the experimental evidence serves to confirm the reasonable hypothesis that the laws of physics are constant and are not changing in time.

This, again, means that our description of nature, that is, the equations in physics, must also possess the same symmetry. The equations themselves are invariant under translations in time. They will involve some arbitrary time, t, and may also involve particular things that happen at different times, t_1, t_2, ... For example, I may drop a ball off the Leaning Tower of Pisa at the instant $t_1 = 9{:}00{:}00$ AM, and I want to calculate how far the ball has fallen a second later, at the time $t_2 = 9{:}00{:}01$ AM. But in any correct equation that describes the time evolution, we can replace all times appearing by new times that are shifted by any constant amount. That is, we can equally well use $t + T$, $t_1 + T$, and $t_2 + T$. The quantity T cancels out of any equation I use to describe the motion, like the translation D in our previous example of translations in space. If I choose $T = 3$ hours, then my physics problem determines where the ball is at 12:00:01 PM, when it was dropped at exactly 12 noon. The result for the height of the falling ball at any subsequent time will be exactly the same, because the laws of physics are time-translationally invariant.[5]

We have seen that energy conservation is a consequence of the unchanging of the laws of nature with time—the heart of Noether's theorem—and we can now turn the argument around and use the observation of energy conservation to conclude that the laws of physics are not changing. We find that the laws of physics must not be changing locally, over extremely short timescales, even time intervals as small as 1/10,000,000,000,000,000,000,000,000,000 (10^{-28}) seconds! By indirect constraints from very rare processes involving the decays of heavy quarks (very tiny particles that we'll discuss in later chapters), we can infer that the constancy of the laws of physics is valid on even somewhat shorter timescales than this.

ROTATIONS

Consider a wine bottle with its label removed. We see that performing the *transformation* of *rotating* the wine bottle around the vertical, about its "axis of symmetry," produces no apparent change in the physical appearance of the bottle. We can photograph the bottle before and after the transformation and find that no discernable change has taken place. There can be other objects in the picture, such as a wheel of cheese, or a bowl of fruit, yet if we have rotated the bottle carefully about its symmetry axis the scene appears unchanged (see fig. 5).

The axis of symmetry is an imaginary line in space that runs through the bottle from the center of the base, through the cork in the mouth of the bottle. We imagine that the symmetry axis is held fixed in space when we perform the rotation. Note that it is important that the label of the bottle has been removed, since the label is an indicator that notably changes position when we do the rotation.

The appearance of the wine bottle is unchanged by the act of rotation through any angle we choose. The symmetry goes well beyond appearances—any physical system, the atoms of glass, the wine that may be left in the bottle, or the cork in the mouth of the bottle, does not change its *physical properties* in any way when we rotate the system. This is more than a symmetry of appearance—it is a symmetry of physics as well. Space itself has no preferred orientation. The laws of physics don't know the difference between "up" and "down," "forward" and "backward," or "sideways."

Gedankenlab discovered that space has continuous rotational symmetries of the laws of physics. Space has the same symmetry as the full rotational symmetry of a perfect three-dimensional sphere. A sphere (or a spherical system) can be rotated about any axis that passes through its center. The rotation angle can be anything we want it to be, so let's take it to be sixty-three degrees. After this rotation (again, an "operation" or "transformation"), the appearance of the sphere is not changed. Again we say that the sphere is "invariant" under the "transformation" of rotating it about the axis by sixty-three degrees. Any mathematical description we use of the sphere will also be unchanged (invariant) under this rotation.

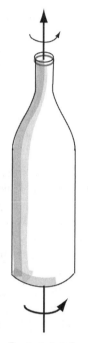

Figure 5. A label-free wine bottle can be rotated through any finite amount, or "continuously," about its symmetry axis without a change in its appearance or physical properties. Alternatively, we can revolve ourselves around the bottle with no change in its appearance or physical properties.

There are an infinite number of symmetry operations (rotations) that we can perform upon the sphere. Furthermore, there is no smallest nonzero rotation that we can perform; we can keep on endlessly performing smaller and smaller, or more "infinitesimal," rotations of the sphere. Thus we say that the symmetry of the sphere is continuous.

The rotational symmetry of the laws of physics is a *continuous symmetry* since the rotation angles can be anything we choose. Clearly, there are an infinite number of symmetry operations that we can perform upon a circle, or a sphere. Again, there is no smallest nonzero rotation that we can perform. We thus say that the symmetry of the circle or sphere is continuous. We say that the sphere or cylinder is *invariant* under the *transformation* of rotating it about any axis passing through its center by 56.54862 . . . degrees, or $\pi/10$ radians, or any other angle we choose. By contrast, a three-bladed propeller or an equilateral triangle appears the same only if the rotation is performed through exactly 120, 240, or 360 degrees, thus providing an example of a *discrete symmetry*. Discrete symmetries have minimum nonzero steps—hence "discrete" symmetry operations. A continuous symmetry, with its infinite number of symmetry operations, is a "bigger" symmetry than a discrete symmetry. Hence, continuous symmetries are very powerful constraints on the structure of space and time. It turns out that it is actually mathematically easier to analyze continuous symmetries, because the powerful techniques of differential calculus can be brought to bear, whereas discrete symmetries pose many challenging counting problems in their analysis.

The laws of physics, therefore, do not depend on how a laboratory is oriented in space. Alternatively, we can actually *revolve ourselves,* like planets going around the Sun, around the sphere, held fixed in space, and the sphere appears to be physically the same. The rotational symmetry of a spherical object is therefore intimately connected to the more general rotational symmetry of space itself. We can't actually distinguish between rotating a spherical object and rotating the entire universe about the spherical object!

We could test this, at least in principle, through experiments done at Fermilab (as opposed to Gedankenlab). If we measure some physical quantity, such as the exact way that a "neutral K-meson" (a particularly interesting elementary particle that we'll encounter again later) decays as it travels in a certain direction in space, we can in principle check whether or not we get the same result at 12 noon as we do at 6 PM. We have to be very careful to make sure that there are no systematic effects in our experimental apparatus when we do this, however. We must make sure, for example, that the voltage on the line coming into the detector doesn't change enough to throw our measurement off, let's say, between 12 noon and 6 PM, when people in the surrounding communities are returning home from work and turning on their air conditioners or using their microwave ovens to cook dinner, and so on (recall that the Acme Power Company was fooled by the *g*-meter fluctuating due to air raid siren tests in the nearby communities—we wish to avoid being fooled by such things ourselves when we perform our physics experiments).

Between 12 noon and 6 PM Earth rotates through an angle of ninety degrees in space. The laboratory, due to its latitude, rotates through a somewhat smaller angle (if the latitude were exactly forty-five degrees from the equator, the lab would only rotate through sixty degrees in space between noon and 6 PM), but the lab *is* rotated nonetheless. We can thus check our data and see that the K-meson behavior is no different at noon than at 6 PM. Of course, this is checking for time dependence as well as orientation dependence, but since we get an answer that is exactly the same, we see that there is likely no effect either of orientation or of time dependence. The main point is that a neutral K-meson's behavior doesn't depend in any way upon how the K-meson, or the lab apparatus that measures it, is oriented in space. The laws of physics are rotationally symmetrical.

Again, the mathematical description of a physical quantity with rotational invariance must itself be symmetrical. As a simple example, we can consider the length of our classroom pointer. We assume that the pointer is lying on a table, and the handle is fixed to lie at a particular point. The tip of the handle is located at some other point on the surface of the table. Since the surface of this table is two-dimensional, we must consider a two-dimensional coordinate system. Then the tip is located at (x,y), and the handle we choose to be located at the origin, or the point $(0,0)$—we can always use translational invariance to place the handle at the origin of the coordinate system. The formula for the length of the pointer, L, is given by the Pythagorean theorem, $L^2 = x^2 + y^2$ ("the square of the length of the hypotenuse equals the sums of the squares of the other two sides"—or did the Scarecrow first say this to the Wizard of Oz?).

Suppose we now rotate the pointer, through any angle, θ, we choose, holding the handle fixed. The tip is now located at the new point (x',y'), while the handle remains at the point $(0,0)$. We can actually use a little trigonometry to write the new (primed) coordinates in terms of the original, unprimed ones. Clearly, the tip ends up somewhere on the circumference of the circle of radius L with the center at the origin $(0,0)$. We find for the length after the rotation, $L^2 = x'^2 + y'^2$. The result does not depend upon the angle θ, and our formula (the Pythagorean theorem) for the length of the pointer is *invariant under rotations*! That is, the same formula for the length applies before and after a rotation and is independent of the rotation angle, so the formula itself has rotational symmetry.[6]

THE SYMMETRY OF MOTION

Gedankenlab discovered one more, and extremely profound, symmetry about space and time: *the outcome of any measurement of the basic parameters of physics does not depend upon the state of motion of the lab, when the lab is uniformly moving through space with any velocity.* If the lab were not uniformly moving with a fixed velocity, then it would be accelerating or spinning and would have experienced a number of weird and apparently *fictitious* forces, such as *centrifugal force* (later we'll see that centrifugal force is not really a force but rather the tendency for something to continue moving in a straight line). So, the lab management restricted this statement to "uniform motion," that is, motion with a constant, fixed velocity. They didn't have to do this, but it kept the bookkeeping simpler. More generally, this is called the *principle of relativity*. It is intimately related to the existence of something we'll encounter shortly, called *inertia*, and forms the basis of Einstein's special theory of relativity.

Not all motion is uniform, however. One day Gedankenlab got dangerously close to a supermassive black hole (see fig. 6). The engines on the lab conked out, and the lab started free-falling into the black hole. No one inside the lab noticed the effect of the black hole, at first, because in free fall there are no "gravitational," or centrifugal, forces—everyone was weightless. They felt as though they were floating in free space, as if there were no black hole nearby about to engulf them.[7] This is why weightlessness can be simulated in an aircraft not far above the surface of Earth—the aircraft can fly on a trajectory that corresponds to free fall, and the astronaut trainees inside the aircraft feel no effect of gravity.

During the free fall toward the black hole, the Gedankenlab experiments obtained the same measured values of all the basic parameters of physics as they did when the lab was in uniform motion at a constant velocity, way out in empty space. Fortunately, someone finally looked out the window and noticed they were minutes away from hitting the event horizon of the black hole, from which they would never be able to exit. They managed to switch on the emergency thrusters, and Gedankenlab made a narrow escape.

During the thrilling thruster escape, Gedankenlab accelerated away from the black hole at 3 *g*'s, making everyone feel as though they weighed three times their normal weight on Earth (much worse than the feeling after a fine Thanksgiving dinner). Nonetheless, during the acceleration away from the black hole, all of the experiments still recorded the same values for the fundamental parameters of physics (even though there were some technical glitches having to do with wires breaking loose and equipment that wasn't bolted down falling and smashing on the floor).

The fact that the laws of physics are the same in a state of free fall in a gravitational field, which includes uniform motion in the absence of gravity, is a profound enhancement of the idea of the symmetry of motion. This symmetry underlies Einstein's general theory of relativity. The laws of physics can be formulated in such a way that they are independent of the general state of motion of the observer. This is the deep symmetry of motion: *we feel gravity only when we are not in free fall.* Acceleration is intimately related to the sensation of gravity.

It is easiest to focus on the simpler case of uniform (constant-velocity) motion. Uniform motion is also a continuous symmetry of the laws of physics, because we can move at any velocity, can vary the velocity to a different one (by acceleration) continuously, and can observe

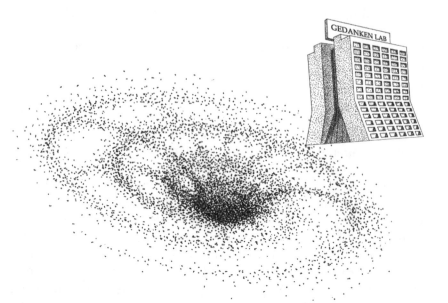

Figure 6. The day that Gedankenlab nearly fell into the black hole Tartarus. (Illustration by Shea Ferrell.)

the same laws of physics at work. So, viewed as a symmetry, it is this *change in the velocity of a system* that is the transformation or symmetry operation under which the laws of physics remain unchanged. Like the term *rotation*, which is the symmetry operation that changes orientation, we call a change in a state of motion of a system a *boost*. Hence, the laws of physics are invariant under boosts. We'll see later that boosts can be viewed as "rotations" in the four dimensions of space and time. This description of motion occurs in Einstein's special relativity, where the symmetry of motion is therefore an extension of the idea of the rotational symmetry in space.

Though the invariance of physics under boosts is called the principle of relativity, which we usually associate with Einstein, the idea actually began with Galileo, who was the first to comprehend inertia, the tendency of objects to move with a uniform motion unless acted upon by a force. This was the greatest conceptual leap of the human understanding of nature and marked the true beginning of the science of physics. Relativity and inertia became significantly refined by Einstein, who was driven by the remarkable properties of light and electromagnetism. It is probably safe to say that the principle of relativity and its equivalent, the principle of inertia, form the cornerstones of all of physics. We'll examine these principles in greater detail in chapter 6.

"GLOBAL" VERSUS "LOCAL"

A subtle question keeps arising in this discussion: are symmetries the same at the shortest times and distances as they are at the largest ones? Perhaps the laws of physics only appear to be constant when examined over very long times, such as the age of the universe. Could they be changing rapidly over extremely short timescales, such as the time it takes for light to travel the diameter of an atomic nucleus, or a proton, or even significantly shorter timescales than this? Or might there be symmetries at the shortest distances and shortest instants of time that are not evident in the universe at large? These are really good questions.

Questions concerning the "large-time" or "large-distance" shape and structure of a universe are called *global* questions. These questions involve the *distribution of the matter* in the universe and what caused this distribution to occur, the kinds of questions that a cosmologist would ponder. Is the universe flat, like an infinite chessboard, extending off to infinity in all directions? Or is the universe infinite in one dimension but perhaps finite and circular in another, so that the global universe looks like an enormous tube, or a cylinder? Or is the universe shaped like the surface of a gargantuan ball, a sphere? Or is it shaped like the surface of a humongous doughnut (something called a *torus*; see fig. 7)?

Global questions pertain to the history of the universe and to whatever produced or created it. How did the universe come into being? What determined its size and shape? How will it continue into the future? These questions are in a fundamental way, however, linked to questions about the very shortest distances in the universe. Global questions can be answered in principle, but they may be very difficult to answer in practice.

Particle physicists, on the other hand, generally focus upon the smallest objects in nature and the shortest distances of space. They try to make measurements about their world *locally*. They study the world as it is in their own backyards. The *local* structure of space (and time) is associated with the questions, What are the symmetries at the very smallest distances of space and time? What are the fundamental constituents of

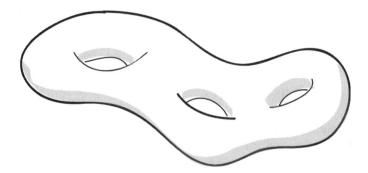

Figure 7. A two-dimensional universe might have the shape of the *surface* of an extended doughnut with many holes. The *local physics* can be the same as at any point on the surface of a sphere; yet the *global physics* is distinguished from that of a sphere by the doughnut topology, with a finite number of doughnut holes.

matter? What are the basic forces through which matter interacts? These questions address the internal fabric of space and time, perhaps the materials and glue that hold it all together, and the fundamental laws of nature.

The conceptual differences between local and global aspects of the universe can be understood in terms of a large soap bubble of the sort blown by children. The larger soap bubbles are made by dipping a metal loop into a pail of liquid soap, and then allowing the wind to blow through the loop, forming a grand and graceful bubble. The bubble forms, having a gracefully transparent and undulating shape, with a tinge of rainbow color. It then oscillates a little bit before it finally settles into the shape of (globally) a sphere. At short distances (locally) it is composed of "goo," that is, soap. On larger distance scales we enter the realm of global physics.

Global questions about the bubble deal with the size and shape and the undulating properties of the overall soap bubble universe, such as, How big a bubble can we blow? The short-distance (local) questions deal with the composition of the soap itself, such as, What is soap and what is it made of? Why is the soap transparent, and gooey, allowing such large bubbles to form? Clearly the local questions are extremely relevant to the existence and properties of the soap bubble. For example, if the soap is diluted with too much water the resulting soap bubbles get smaller and smaller. Too little water and the bubbles won't form at all. The size of the soap bubble universe tells us something about soap itself. The detailed structure of soap, at the shortest distances, goes by the following fancy technical terminology: *molecules forming an anionic surfactant composed mostly of alkali (sodium or potassium) salts of triglycerides (fatty acids)*. This is an interesting and complex science all its own.

At short distances we enter an entirely new and different set of questions. We can ask countless questions about the molecular arrangement of soap, how it forms, and how it can be modified. Understanding the laws of the molecular composition of soap is potentially very useful. Through them we can invent new soaps that keep our things clean! We can invent biodegradable soaps that don't pollute and that can be used to clean up gigantic coastal oil spills, or maybe we can invent new soaps that are better lubricants. How about soapy glue, or soapy machine oil, or magnetic soap, or nanotechnology soap that cleans itself up after a spill?

The local laws of nature are fundamental and all-pervasive. The local laws determine ultimately what can or cannot exist. The global universe is ultimately one of the many gadgets or inventions or *applications* that one can make from the detailed understanding of the local laws of nature.

Beyond the space-time symmetries we have just discussed, there are continuous symmetries that apply not to space and time but to matter and the intrinsic properties of matter that are described by quantum mechanics, as well as the properties of the elementary particles. The local and global aspects of these symmetries are equally profound. These symmetries lead us to the concept of *charge* and to the fundamental forces in nature. We'll describe this in later chapters, but for now we move to the implications of the continuous symmetries of space and time and their consequences for the behavior of physical systems, as embodied in the theorem of Emmy Noether.

chapter 5

NOETHER'S THEOREM

For every continuous symmetry of the laws of physics, there must exist a conservation law.

For every conservation law, there must exist a continuous symmetry.

—Noether's theorem

CONSERVATION LAWS IN ELEMENTARY PHYSICS

Noether's theorem is the deepest and most direct connection we have between dynamics—forces and motion and the fundamental laws of nature—and the abstract world of symmetry. The theorem was proven in 1915 by the young Emmy Noether shortly after her arrival at the University of Göttingen.

The theorem provides us with a connection between continuous symmetries of the laws of physics and the existence of corresponding conservation laws. A conservation law is the statement that there exists a physically measurable quantity (such as the total energy of a system) that *doesn't change* in any physical process (e.g., the total energy is always the same before and after any process occurs). Such a physical quantity is called a *conserved quantity*. Noether's theorem unifies the concepts of symmetry and conservation laws and tells us how symmetries are most directly manifested in nature.

We'll now focus upon the *space-time conservation laws*. These are the particular conservation laws that follow from the symmetries of translations and rotations in space and time that we described in the previous chapter. They lead to the conservation laws of *energy, momentum,* and *angular momentum*. These are usually studied in high school physics classes, where the experimental evidence for the conservation laws is stressed. Unfortunately, however, in high school physics classes the profound connection to the symmetries of the laws of nature, through Noether's theorem, is never, or at most rarely, mentioned. Through the theorem, however, the conservation laws actually become easier to comprehend—they are "demystified" when viewed as a consequence of symmetry.

For now we will merely state Noether's theorem as a fact, without giving its mathematical proof (we'll see how it works as we proceed). Noether's theorem applies to all of physics—to both classical (with or without special relativity) and quantum mechanics, though the concept of "observable" has to be refined in the latter case.[1] Indeed, there are many conservation laws in physics, in addition to the space-time conservation laws we'll be exploring here. These include such things as the conservation of electric charge; the total number of *baryons* (protons plus neutrons minus antiprotons and antineutrons) in a system; the total number of leptons, such as electrons and electron-neutrinos; the *quark color* of a state, such as the proton, containing quarks and gluons; and so on. Each of these conserved quantities—charge, baryon number, electron number, color, and so forth—comes from a continuous symmetry lurking deep within the structure of the laws of nature. In fact, as we have stressed earlier, and we'll see more later, the laws of physics themselves are essentially *defined by symmetry principles!*

CONSERVATION OF MOMENTUM

As we have seen, it is an experimental fact of nature that the laws of physics are invariant under spatial translations. This is a strong statement. This is a continuous translational symmetry of the laws of physics. The hypothesis that space is continuously translationally invariant is equivalent to the statement that, from the point of view of the laws of physics, *any* point in space is equivalent to *any other* point in space.

The symmetry is such that translations of any physical system or apparatus, or equivalently, the translation of the coordinate system that we use to describe things, by any amount, in any direction, does not affect the laws of nature governing the system. Therefore, the outcome of any experiment we perform is not affected by translating the whole laboratory somewhere else in space. Put simply, the laws of physics and the equations that express these laws are invariant under translations.

Emmy Noether's theorem, in the case of continuous translational invariance of the laws of physics in space, implies the *law of conservation of momentum*. Ah ha! We learn in a high school physics class that the *total momentum* of an isolated system remains constant in time, no matter how the particles in the system interact. For instance, when two billiard balls collide, the total momentum before the collision exactly equals the total momentum after the collision. Now we see that there is a more fundamental reason for this—the laws of nature are the same everywhere in space! So, let's remind ourselves what momentum is.

According to Noether's theorem, in a three-dimensional universe, such as the one we live in, there are three *perpendicular directions* in which we can translate a physical system (scientists call this three *mutually orthogonal* translations). Since the system can be translated in any of the three directions of space, there must be three conserved momenta, *one associated with each direction of space*. The conserved quantities are in one-to-one correspondence with the three perpendicular translational *degrees of freedom* in space. Therefore, momentum, like the position of a particle, or the velocity of a particle, or the force acting upon a particle, has both a *direction in space* and a *magnitude*. Such a thing is called a *vector*.[2]

Velocity, for example, is a vector. It is the measure of the motion for any object, and it clearly has a direction, the *direction of motion* of the object, and a magnitude, which is the *speed* of the object. Technically, therefore, speed is just a numerical quantity and implies no direction; I can readily say I am traveling at sixty miles per hour without telling you to which point on the compass I am headed. To tell you my velocity vector, I must tell you my speed and my direction: "I am going sixty miles per hour due north."[3]

We often graphically represent a vector by an arrow pointing in the direction of the vector, with a length that represents its magnitude. For low-speed objects, such as a tortoise, we would draw a vector as an arrow pointing in the tortoise's direction of motion; the arrow would have a short length, reflecting the tortoise's low speed. For a hare, we would similarly draw an arrow for the direction but one with a greater length for the arrow, representing the hare's higher speed.

In Newtonian physics, momentum is the product of the *mass* of an object (a number, which has only a magnitude and no direction) with its *velocity* vector. Momentum therefore has a direction, the direction of motion as determined by the direction of the velocity, and it has a magnitude, the product of the mass and the speed of the object. Momentum is therefore, and indeed must be, a vector. We therefore write the following equation for the object's momentum: $\vec{P} = m\vec{v}$ where m is the mass and $\vec{v}$ is the vector velocity. Recall that the mass, m, is a measure of the quantity of matter of an object but makes no reference to the motion of the object. Velocity, $\vec{v}$, is a measure of the motion of the object that makes no reference to its mass. Momentum is therefore a measure of *physical motion* that includes both the mass and the velocity together. An extremely massive object moving slowly can have just as much momentum as a small object moving quickly. For example, if we think of our tortoise-and-hare example, although the hare has a much larger magnitude of its velocity than the tortoise, the tortoise could have a comparable or much larger momentum than the hare if its mass is much larger than the hare's mass.

We emphasize here that it is the *total* momentum of a physical system that is conserved and not the individual momenta of the parts of the system (*momenta* is the plural of *momentum*). This is true because, when we do a translation in space, we translate the *whole system*, not just one of its parts.

The simplest example of momentum conservation occurs in the radioactive decay of a particle, A, into two fragments, or "daughter particles," B and C. If the parent particle A is initially at rest (zero velocity) in the laboratory, then the initial momentum of the "system" is zero. After the decay happens, the two particles B and C speed off in precisely opposite directions (we say "back-to-back"). The momenta of the two daughter particles must, by conservation of momentum, add to zero, so they must be exactly equal in magnitude but opposite in direction, or $\vec{p}_B = \vec{p}_C$. This result is crucially useful when we consider a more complicated situation, a decay into three particles. In fact, a neutron, one of the constituents of the atomic nucleus, decays into three particles

$n^0 \rightarrow p^+ + e^- + \bar{\nu}$, a proton, an electron, and an (anti-)neutrino.[4] Each of the three outgoing particles has its own momenta $\vec{p}_p$, $\vec{p}_e$, and $\vec{p}_{\bar{\nu}}$, and again these must add to zero.

If a neutron at rest in the lab decays, the produced proton and electron are fairly easy to detect and to track with a particle detector. On the other hand, the neutrino is *extremely* difficult to detect. But, if one observes the proton and electron moving at any angle other than exactly back-to-back—at an angle different than 180 degrees between them— then the law of conservation of momentum forces us to conclude that a third particle, the (anti-)neutrino, must also be involved (see fig. 8). We thus detect the neutrino indirectly, through the law of conservation of momentum. This was the crucial early evidence for the existence of neutrinos.

Another familiar example of momentum conservation is the collision of two pointlike massive objects, such as billiard balls. The objects are designated as 1 and 2 and have, respectively, masses m_1 and m_2 and velocities $\vec{v}_1$ and $\vec{v}_2$. For example, object 1 could be the number 1 ball on a pool table, while object 2 is the number 2 ball. Let us suppose the balls collide. Initially they have a total momentum of $m_1\vec{v}_1 + m_2\vec{v}_2$. After the collision, the velocities will generally have changed, with new velocities $\vec{v}_1'$ and $\vec{v}_2'$, because of the forces and dynamics of the collision. However, at least for billiard balls, the masses do not change (very much). The postcollision total momentum is therefore $m_1\vec{v}_1' + m_2\vec{v}_2'$. The law of conservation of momentum states that $m_1\vec{v}_1 + m_2\vec{v}_2 = m_1\vec{v}_1' + m_2\vec{v}_2'$.

Actually, the collision of two billiard balls, when described at the atomic level, is a very complex thing, involving the interactions of trillions and trillions of atoms. In a collision there is a slight rearrangement of the material itself, some atoms breaking off to become dust, others being squeezed together. The specific arrangement of the atomic positions is caused to vibrate, leading to the emission of the "clacking" sound of the two balls in collision. The entire physical atomic structure of each billiard ball then recoils and spins, rolling in different directions. After the collision, the masses of the billiard balls are approximately the same. This is a good approximation for billiard balls, but it need not be true in general; the objects could change their masses in the process of the collision, as often happens when elementary particles collide and convert to different elementary particles.

Therefore, at a very detailed microscopic level, the total momentum is the sum of all the individual momenta of all the atoms in both billiard

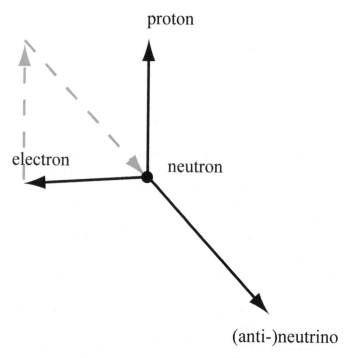

proton

electron   neutron

(anti-)neutrino

Figure 8. A neutron with zero initial momentum decays into a proton, an electron, and an (anti-)neutrino. The three momentum vectors of the outgoing particles are drawn (solid lines). The vectors "add to zero," which means (graphically) if one were to walk along one of the vectors to its end, such as that of the electron, then turn and walk along parallel to another to its end, such as the proton (dashed line), then turn and walk parallel to the third, such as the (anti-) neutrino (dashed line), one would return to the origin.

balls at some initial instant of time. However, for the billiard ball collision, our simplified "two-body" description, in which the total momentum is just $m_1\vec{v}_1 + m_2\vec{v}_2$, is a very good approximate description. Indeed, we don't get very far in physics if we cannot make such approximations to an otherwise complex and unwieldy situation. Much of the art of physics is knowing what approximation to make. So, although it is the total momentum that must be conserved in the collision, for two billiard balls this is just the sum of the momentum vectors of each billiard ball before and after the collision, to a very good approximation.

When would this be a bad approximation? Object 1 could be the planet Earth and object 2 an enormous asteroid named Zlot, an asteroid about the size of the Moon. One can gain some appreciation of this by imagining the horrendous experience of living on Earth if it were struck in a grand collision by the asteroid Zlot, and the complex phenomena that would ensue. Zlot and Earth need not come into contact; in fact, they could approach and "touch" each other through the force of gravity, even though they would remain many thousands of miles apart. This would still have a very unpleasant, indeed nonsurvivable, effect for the residents of planet Earth (or Zlot), as humongous mountains and tides would be upheaved and enormous geological shockwaves would engulf the entire planet, reshaping and deforming the planet's entire surface. Ocean and ground waves hundreds of miles high would result, and the two planets might shatter into billions of pieces! Much of the debris would end up becoming reamalgamated into the newly rearranged Earth and Zlot, but a great deal of it would fly out into space and reclump to make new and smaller asteroids and meteorites, many of which would rain down on the newly formed world(s) for centuries.

Despite the complexity of the full set of physical processes that would ensue in this unthinkable collision, the conservation law of momentum assures us that the *total momentum* of the total physical system of Earth and Zlot must remain the same, before and after the collision. For Earth and Zlot, the initial total momentum would be $m_{\text{Earth}} \vec{v}_{\text{Earth}}$ + $m_{\text{Zlot}} \vec{v}_{\text{Zlot}}$. The final total momentum would be $m_1' \vec{v}_1' + m_2' \vec{v}_2' + m_3' \vec{v}_3' +$ . . . , where we have to add up all the momenta of all the outgoing debris with the various masses and velocities of the different bits, chunks, and pieces of debris. Despite this calamitous catastrophe, in which everything we ever knew would be literally wiped off the face of Earth, one simple fact remains: $m_{\text{Earth}} \vec{v}_{\text{Earth}} + m_{\text{Zlot}} \vec{v}_{\text{Zlot}} = m_1' \vec{v}_1' + m_2' \vec{v}_2' + m_3' \vec{v}_3' +$ . . .—the total momentum of the collision would be conserved! It may not be much, but at least that's something to hold on to.[5]

Momentum conservation cuts across the complexity of a physical process and always remains valid, no matter what. Another example is the explosion of an artillery shell in midair, in which thousands of shards of shrapnel from the explosion each have their own momentum, but the sum total is exactly that of the initial momentum of the shell itself.

The conservation of momentum is a powerful constraint on what can or cannot happen in any physical process, no matter how complex and no matter what forces of nature are involved. You may well ask, "Isn't Earth's momentum always changing and *not* conserved?" Indeed, Earth is going around the Sun as it moves in its orbit and is continuously changing its velocity (by changing the direction of motion the velocity vector changes, even though the speed remains the same). The *total momentum*, however, must still be conserved in this process, but now we must broaden the definition of "system" to include the Sun. The Sun is pulling on Earth to change its velocity and thus to change its momentum. But Earth, on the other hand, is also pulling back on the Sun and is changing its velocity too, but by a very tiny amount. The orbiting planets actually produce a slight "wobble" in the motion of the Sun.[6]

In fact, the discovery of new planets orbiting distant stars has recently been achieved by a technique nicknamed "wobble watching." Astronomers have detected wobbles in the motion of stars that presumably have supermassive planets like Jupiter in close orbits, which maximizes the wobble. There are over fifty of these "exoplanets"—planets that lie outside of own solar system—now known, and the number is increasing.[7] When we were kids we never thought that planets orbiting about other stars would actually be discovered!

Actually, physicists long ago understood that momentum was conserved, long before Noether's theorem. It is encoded into Newton's laws of motion and probably was discovered by Newton himself. If we apply a force $\vec{F}$ (a vector) on an object of mass m for a brief period of time, t, then it turns out that its momentum will change by the amount $\vec{F}t$, and therefore its velocity will change by the amount $\vec{F}t/m$. We call $\vec{F}t$ the *impulse* applied to an object. We see that *impulse equals the change in momentum*, and this evidently explains why the starship *Enterprise* of *Star Trek* fame is equipped with "impulse engines."

Newton realized that when object 1 collides with object 2 there is a force that object 1 exerts upon object 2, which we call $\vec{F}_{12}$. Likewise, there will be a *reactive force* exerted *back upon* object 1 *by* object 2, which is $\vec{F}_{21}$. For example, as baseball player Alex Rodriguez's bat hits the baseball, there is a force exerted by the bat upon the ball, $\vec{F}_{12}$, and a force exerted back, by the ball, upon the bat, $\vec{F}_{21}$. Newton's third law of motion states that these forces must be equal in magnitude but opposite in direction: $\vec{F}_{12} = -\vec{F}_{21}$. Note that this is a *vector equation* since the forces are necessarily vectors, such as acceleration, velocity, and momentum. We therefore see that the change in the momentum of the ball, when it is struck by the bat, is the impulse $\vec{F}_{12}t$, where t is the brief time interval over which the collision takes place. Likewise, the change in the momentum of the bat is the impulse $\vec{F}_{21}t$ in the collision, but this equals $-\vec{F}_{12}t$, from Newton's third law of motion. Therefore, the net change in the *total momentum* of bat plus ball is $\vec{F}_{12}t + \vec{F}_{21}t = 0$. The total momentum is conserved, as it must be, whether in a billiard ball collision or any other collision.

Since big objects are just sums of many small individual components, and we can more or less think of all the interactions as analyzed into many two-body interactions, it follows that the total momentum is always conserved for all systems. So, momentum conservation really follows from Newton's third law of motion. But where does Newton's third law come from? Noether's theorem is the deeper statement, implying that the total momentum is conserved, *because* the interactions are determined by laws that don't depend upon where the system is located in space! So, Newton's third law, $\vec{F}_{12} = -\vec{F}_{21}$, also follows from Noether's theorem—and therefore from translational symmetry of the laws of physics! Thus we are beginning to see that the "laws of physics" are really one and the same as symmetry.

We can turn this around. The validity of the law of conservation of momentum is an observational fact and can be tested directly for any process in the laboratory. Noether's theorem, however, implies that space must possess translational symmetry, since we observe momentum conservation in the lab. We can readily test momentum conservation in collisions of elementary particles that happen during very short times, and we still see that momentum conservation always works. This implies that even at very short distances, about $1/100,000,000,000,000,000,000$ inches (10^{-19} m), the translational symmetry of space is still a valid symmetry.

CONSERVATION OF ENERGY

That the laws of physics are invariant under translations in time is a continuous symmetry. What conservation law would then follow by Noether's theorem? As we have seen, it is nothing less than the law of conservation of energy. Since the constancy of the total energy of any system is extremely well tested experimentally, in experiments with billiard balls, planetary orbits, and quarks, the theorem tells us that nature's laws must be invariant under time translations. Conversely, evidence such as from the Oklo fossil nuclear reactors strongly favors the hypothesis that the laws of physics are not changing in time, and therefore Noether's theorem implies that the total energy of any system must be conserved. We should therefore never invest in perpetual-motion and energy-from-nothing schemes, such as the one involving the Acme Power Company.

How does confidence in scientific conclusions arise? For example, how convincing is the consistency of the law of energy conservation, throughout all of the multitude of processes in physics? If one chink in its armor were found, by virtue of its relationship to the fundamental symmetry of time invariance of the laws of physics through Noether's theorem, the whole logic of physics would come crashing down.

In 1898 Marie and Pierre Curie, along with Henri Becquerel, performed the first studies of natural radiation emanating from material substances. At that time the atomic structure, in particular that of the atomic nucleus, was still unknown. These scientists observed the basic forms of naturally occurring radiation typically emitted by unstable atomic nuclei. This radiation was observed to take three different forms, which were classified as *alpha*, *beta*, and *gamma* radiation.

Today we understand *alpha rays* to be the emission of an entire nucleus of helium (an alpha particle) from the spontaneous disintegration of a much heavier nucleus. *Beta rays* are the emission of ordinary electrons (or their antiparticles, positrons) in a nuclear disintegration. *Gamma rays* are very energetic photons, the particles of light, the "quantum of electromagnetism," also emitted by unstable nuclei. Through the detailed study of these "rays," all of the usual laws of physics, such as energy and momentum conservation, were found to be verified for both alpha and gamma radiation; but in the special case of beta radiation, the physicists discovered a disturbing result—it seemed that when an atomic nucleus undergoes the emission of beta rays, also known as *beta decay*, the conservation of energy (and momentum) was apparently *violated*!

The simplest example of beta decay occurs with a single *neutron*, one of the particles found in the atomic nucleus, when it floats about freely through space. The problem, from countless observations, was that the electron and proton energies always added up to something less than the original neutron energy. The momenta of the outgoing electron and proton didn't add up to that of the neutron either, since neutrons at rest in the lab decayed into protons and electrons that were not seen to be emitted back-to-back. There thus appeared to be a *missing amount* of energy and momentum in the decay of a neutron. Essentially all beta decays of nuclei are a more complex variation on this process, where the neutron is typically bound within the nucleus.

This missing energy and momentum in beta decay was a considerable mystery to physicists for many years. Niels Bohr, one of the founding fathers of quantum mechanics, attempted to explain this phenomenon with the hypothesis that the conservation of energy and momentum has only a limited validity in the world and that the beta decay processes were exhibiting, for the first time, a true violation of these conservation laws. Bohr, a brilliant and creative thinker, had already seen in the early part of the twentieth century that our detailed understanding of energy and momentum was significantly modified by the rules of quantum mechanics, and perhaps this was an indicator of deeper novelties and surprises yet to come.

Yet his proposal would have shocking consequences. By Noether's theorem, this would imply that, somehow, for the beta decay reaction, the symmetries of continuous translational invariance in space and time don't hold. One could imagine, instead, that space and time do form a kind of crystal lattice, in which continuous translational symmetry wouldn't really hold in space (and time) after all. This would be a truly astounding discovery—our universe would then be similar to an infinite discrete chessboard. If the law of energy conservation could be violated, then perhaps the idea of an Acme Power Company would not be so far-fetched after all!

Wolfgang Pauli, a young and brash theoretical physicist, could not accept Bohr's idea. The principles of the conservation of energy and momentum up to this point had proven valid in all domains of physics. It seemed unnatural to Pauli that the violations would show up *only* in beta decay reactions, where it is apparently seen to be a very large effect, and not show up elsewhere. Everything is interconnected at some level in physics, so why, if this were true, wouldn't there be small violations of energy and momentum conservation detected in all other processes, especially given that the underlying symmetries of space and time enforce energy and momentum conservation? Wouldn't any violation of these precious symmetries be universal, felt by all forces in nature, and not just a property of beta decay? It made no sense to Pauli.

Therefore, in 1930 Pauli postulated the existence of a new and unseen elementary particle that he proposed was also produced, together with the proton and electron, in the beta decay reaction. This new particle would carry no electric charge and would therefore escape the decay region totally unobserved. This unobserved particle would be carrying off the missing energy and momentum and would thus be preserving the validity of the conservation laws. In other words, physicists could compute the missing energy and missing momentum required to maintain the conservation laws in any beta decay reaction, and this would be the exact momentum and energy carried off by the new particle. In Pauli's own words, written December 4, 1930, in a response to an invitation to attend a conference on radioactivity,

> Dear Radioactive Ladies and Gentlemen,
>
> As the bearer of these lines, to whom I graciously ask you to listen, will explain to you in more detail, how because of the "wrong" statistics of the N and Li^6 nuclei and the continuous beta spectrum, I have hit upon a desperate remedy to save the ... law of conservation of energy. Namely, the possibility that there could exist ... electrically neutral particles, that I wish to call [neutrinos], which have spin 1/2 and obey the exclusion principle ... and in any event [have masses] not larger than 0.01 proton masses. The continuous beta spectrum would then become understandable by the assumption that in beta decay a [neutrino] is emitted in addition to the electron such that the sum of the energies of the neutron and the electron is constant. ...
>
> I agree that my remedy could seem incredible because one should have seen these [neutrinos] much earlier if they really exist. But only the one who dares can win and the difficult situation, due to the continuous structure of the beta spectrum, is lighted by a remark of my honoured predecessor, Mr Debye, who told me recently in Brussels: "Oh, it's well better not to think about this at all, like new taxes." From now on, every solution to the issue must be discussed. Thus, dear radioactive people, look and judge.
>
> Unfortunately, I cannot appear in Tübingen personally since I am indispensable here in Zurich because of a ball on the night of 6/7 December. With my best regards to you, and also to Mr. Back.
>
> > Your humble servant,
> > W. Pauli[8]

(One notes that, whatever else can be said about Pauli, his priorities were rock solid!)

This particle is now called the *neutrino*. Therefore, when the neutron decays in free space, it produces a proton, an electron, and an (anti-)neutrino. In our modern parlance, the electron is produced together with the anti-electron-neutrino. The sums of the final energies and momenta will be exactly the same as the initial energy and momentum of the original parent neutron. Notice also that the neutrino, with zero electric charge, allows the beta decay reaction to satisfy the law of conservation of *electric charge*. The zero electric charge of the neutrino means that it can't be easily detected—it lacks the "handle" of electric charge that we can grab onto through electromagnetic fields in our particle detectors.

Pauli was right! Neutrinos do indeed exist. Neutrinos were ultimately directly detected by Clyde Cowan and Fredrick Reines in 1956. These neutrinos were emitted from the decays of neutrons present in nuclear fission processes in the cores of reactors at nuclear power plants. However, we now know that there are at least three different kinds, or "flavors," of neutrinos. Leon Lederman (one of the authors of the present book), Mel Schwartz, and Jack Steinberger demonstrated in 1962 that neutrinos are produced with distinct identities, by detecting the muon-neutrino, which is a different particle than the electron-neutrino. Today we know that there are three kinds of neutrinos: *electron-neutrinos, muon-neutrinos*, and *tau-neutrinos*. This zoology of particles will be further elaborated in a later chapter—it is rife with symmetry, both exact and approximate.

Today the search is on for processes in which the three neutrinos "oscillate," or change their identity. For example, a muon-neutrino produced in a high-energy collision may switch its identity into a tau-neutrino at a later time. Pauli, by keeping the faith in energy and momentum conservation, had opened the door to a whole new family of elementary particles—the neutrinos. That door is being flung open wider still as we type this paragraph: the subject of neutrinos is one of the hottest research topics in both particle physics and cosmology. We should add that experimentalists still often look for missing energy and momentum in their detectors in particle collisions, but this is always interpreted nowadays as evidence for a new particle, never as evidence for the breakdown of the conservation laws of energy and momentum. Our faith, or should we say confidence (as science is not faith-based), in the symmetries of the structure of space and time, and Noether's theorem would, at this point, be very hard to shake.

As we noted earlier, energy can take many forms. Kinetic energy, the energy associated with motion, is just one form of energy. The problem with measuring total energy conservation in general is that, although kinetic energy is usually easy to observe, it can be converted to other forms of energy that are harder to observe, such as heat, sound, potential energy, the energy of crumpling, and so forth. Furthermore, as we've seen, we can convert energy of motion into potential energy, or vice versa. This tends to hide the direct and obvious effects of energy conservation, making it all seem somewhat enigmatic. A railroad boxcar coasting along a level track has a certain kinetic energy of motion. As it coasts uphill, it will eventually come to rest, having lost kinetic energy to potential energy associated with climbing upward against the pull of gravity. We say that the boxcar has "done work" against the force of gravity, giving up its kinetic energy, which is now stored in the gravitational field. As the boxcar begins to accelerate back downhill, we say that gravity is "doing work" to give up potential energy and give back kinetic energy to the boxcar. It all works out in the end, however—the total energy of a physical system is conserved.

However, in certain special cases we can observe collisions in which the initial *kinetic* energy (just the energy of motion) and final *kinetic* energy are the same, that is, collisions in which the kinetic energy itself has been conserved. There can be no energy lost to deformation or heat or sound in such collisions. These special collisions are called *elastic collisions*. This should not be confused with elastic stretch bands, such as are found in men's briefs. There *elastic* means that the shape or form of the waistband is preserved even after one wears them all day.

A beautiful example of an elastic collision, to a good approximation, is the coffee-table toy consisting of colliding steel balls, in which one ball, attached to a thread, is lifted at the end and allowed to collide with a line of five other balls. Then the ball at the opposite end pops up and swings back down, colliding in the reverse direction with the five other balls. This causes the first ball to pop up, and the process repeats for a long time, testifying to the near-perfect conservation of the kinetic energy, in really good elastic collisions. Steel hitting steel tends to produce a rather good elastic collision, because steel is fairly incompressible, so energy is not wasted in deformation. A little energy is lost to sound, heat, and air motion, so the balls do eventually come to rest. The elasticity (slow loss of kinetic energy) of steel on steel is one reason why railroads are so energy-efficient—a well-oiled boxcar can coast for many miles on a well-built, level track before friction dissipates its kinetic energy.

Note that, while energy is conserved, mass is not necessarily conserved. This often confuses people who are first learning physics, because they are told that mass and energy are equivalent through Einstein's famous formula, $E = mc^2$. In fact, the latter statement is not correct, because massless particles, such as photons, have energy. In addition, the formula changes when particles are moving. Therefore a nucleus or an elementary particle can change to a different nucleus or particle of a different mass, and the total energy of the process is still conserved. (Some other particles are usually involved in the process to conserve energy.)

However, at lower energies, such as in chemical or biological processes—that is, nonnuclear processes—mass *is*, to a very good approximation, conserved. Hence, "the principle of mass conservation," formulated long ago by Archimedes, is often used in chemistry.

CONSERVATION OF ANGULAR MOMENTUM

We also live in a world where the laws of physics are rotationally invariant. According to Noether's theorem, *the conservation law corresponding to rotational symmetry is the law of conservation of angular momentum.* Just as momentum is a measure of physical motion of a system in a straight line, such as the direction of a velocity vector, angular momentum is a measure of rotational motion. Angular momentum pertains to the rotational invariance of the laws of physics and therefore pertains to circular motion as well.

Noether's theorem tells us that angular momentum is related to rotations, just as momentum is related to spatial translations. We actually define a rotation to be a type of vector. To do this, we must use the *right-hand rule.* For example, consider a spinning Frisbee or a gyroscope. The rotation is defined by curling one's fingers, of the right hand, in the direction of rotation. The thumb of the right hand then defines the sense of the rotation. The thumb points in the direction that we call the *axis of rotation,* an imaginary line perpendicular to the *plane of rotation.* The right hand determines the direction along the axis (up or down) in which the rotation is defined to point. With a little practice, you can get used to the idea of the rotation defined as a vector using the right-hand rule. The picture in figure 9 is worth a thousand words.

In the simple case of the planet orbiting a star, the angular momentum is the vector we get by curling our right-hand fingers in the direction of motion of the planet in its orbit and noticing where our thumb points. For a counterclockwise-moving planet, viewed from high above its orbital plane, the angular momentum vector is perpendicular to the plane of the orbit and points out of the plane toward the viewer.

The right-hand rule determines the direction of the angular momentum vector for any spinning or orbiting object, but what then determines the magnitude of this vector? Suppose we have a planet in a circular orbit about a very massive star. The orbit has a radius, R, and at any time the planet has a momentum $\vec{p} = m\vec{v}$, where $\vec{v}$ is the velocity vector. In a circular orbit the velocity vector is always *tangent* to the orbit

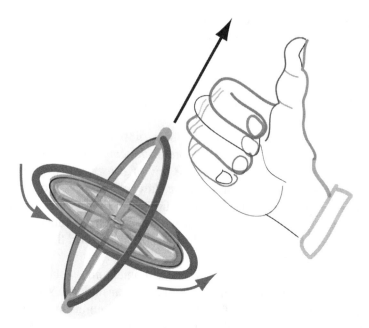

Figure 9. The angular momentum of a gyroscope is defined by the rotation vector, the direction of which is defined by the right-hand rule. (Illustration by CTH.)

(perpendicular to the direction to the center of the orbit and lying in the plane of the orbit). We assume the star is so massive that its wobble motion is negligible. In this situation the magnitude of the angular momentum vector is simply the radius, R, times the magnitude of the momentum (the "length" of the momentum vector), which we write as $m|\vec{v}|$, or the mass times the speed of the planet (in a circular orbit, the speed of the planet never changes). Thus the angular momentum vector in this case has a magnitude of $m|\vec{v}|R$ and has a direction that is perpendicular to the planet's orbit.

Hence, our little orbiting planet as it traverses its orbit about its star, year after year, has a conserved (vector) quantity, a consequence of the fact that the laws of physics do not depend upon orientation of the whole system in space, called angular momentum. Scientists often use the symbol $\vec{J}$ to represent the angular momentum vector. The planet's angular momentum is conserved and cannot change as long as our "system" consists only of the star and the planet. Should the wayfaring asteroid Zlot come by and collide with this little system, the angular momentum of star-plus-planet can be changed, but then the total angular momentum of star-plus-planet-plus-Zlot will be conserved. So, if star-plus-planet is undisturbed, the angular momentum vector is conserved.

The conservation of the planet's orbital angular momentum implies that the motion must *always* remain in the same plane. Otherwise, the vector $\vec{J}$, defined through our right-hand rule, would change its direction. This is one of the most important discoveries of Johannes Kepler, who first correctly codified the laws of motion of the planets. Kepler also discovered that, for any of the solar system's planetary motions, including extremely elliptical orbits such as that of Mars, there is a universal relationship between the time it takes for the planet to complete one full revolution and the size of the orbit. This relationship is a direct consequence of the conserved magnitude of $\vec{J}$. Kepler's empirical laws of planetary motion thus contain the main consequences of angular momentum conservation. They represent the first discovery of a conservation law in physics, one that applies to a dynamical system involving forces and motion.

Angular momentum conservation, of course, applies to any complicated multibody planetary orbits, such as systems of three or more stars bound together in a cluster, provided we add up all of the individual angular momenta of all of the particles in the system. As with momentum, it is the *total* angular momentum of the system that is conserved, and we can write the vector equation $\vec{J} = \vec{J}_1 + \vec{J}_2 + \vec{J}_3 + \ldots$, where $\vec{J}$ is the total angular momentum and $\vec{J}_i$ are the angular momenta of the various components of the system. To actually solve for the orbital motion in complex multibody situations, using pencil and paper, is nearly impossible, and only a few exact solutions are known. To get general results requires the use of computers or drastic simplifications. However, no matter how difficult the multibody problem may be, we are always guaranteed that the total angular momentum is conserved. This is also true for the more general hyperbolic or parabolic trajectories of comets and of the Zlots of this universe, as well as collisions involving galaxies, vehicles, atoms, molecules, or elementary particles involving any of the many forces in nature.

Massive objects can also spin, and they have angular momentum associated with this kind of motion as well. A child's top is a simple object illustrating spin. When any object with an "approximate size" of R (this should be thought of as the radius of the object in the plane of its spinning motion) and a total mass m is spinning, with its outer extremities moving at a speed v, then it will have a magnitude for its spin angular momentum of $|\vec{J}| = kmvR$. The letter k simply stands for a number, like 0.793, that characterizes the shape and internal matter distribution of the object (the astute reader will note that this is just the size, R, times the momentum's magnitude, mv).

With this simple observation, we can now understand how conservation of angular momentum can be demonstrated in a lecture by what is usually called *the three-dumbbell experiment*. The instructor stands on a rotating table, his hands outstretched, with a heavy dumbbell in each hand (guess who is the third dumbbell?). A student starts the lecturer-plus-dumbbell system rotating on the table (see fig. 10*a*). The instructor-plus-dumbbells thus becomes a spinning system. At first he spins slowly; but then he brings his hands (and the dumbbells) closer to his body. His rotation speed, v, is observed to increase substantially (fig. 10*b*). Why?

What must be kept constant here is the total angular momentum, because of momentum conservation, and this has a magnitude that is $|\vec{J}| = kmvR$. By bringing the dumbbells in closer to his body, his "radial size," R, is therefore decreased. But $|\vec{J}| = kmvR$, the angular momentum, must stay the same, so v must increase to compensate the decrease in R.[9]

In this way, figure skaters employ angular momentum to achieve their impressive spins on the ice as well as rotations in midair. And, as the core of a gigantic Titan star collapses in a supernova, it can form a small relic neutron star, which ends up carrying the entire spin angular momentum of the inner core of the collapsing star. The neutron star must therefore spin at a very high rate. Such an object often emits regular pulses of light as it spins, pulses created by its enormous magnetic field sweeping through the surrounding debris. This remarkable object is called a *pulsar*.

Angular momentum conservation, as in the fixed orientation of $\vec{J}$ in space, contributes to the stability of many systems. Gyroscopes are spinning masses that maintain their orientation in space when they are supported in extremely frictionless housings called gimbals. They can be used as navigational aids where information about orientation is relevant. Other angular momentum–stabilized systems are bicycles, where the wheel rotation provides significant angular momentum, which preserves the upright state of the bicycle. Frisbees are stabilized in flight by angular momentum. Rifle bullets and artillery shells are caused to spin by "rifling," or grooves carved into the internal barrel of a gun to make projectiles spin, for a more stable flight. Quarterbacks spin the football to gain stability and hence precision when passing (hopefully) for a touchdown. Earth itself preserves its axis of rotation, pointed toward the pole

Figure 10. The three-dumbbell experiment. In *a*, Professor Peabody holds in his outstretched arms a pair of dumbbells. He begins to rotate ever so slowly. In *b*, Professor Peabody retracts his arms, bringing the dumbbells close to his body. Angular momentum is conserved, causing his angular velocity to increase substantially. (Illustrations by Shea Ferrell.)

star (the star at the end of the handle of the Little Dipper) as it spins through the day-night cycle.

With a few exceptions, each of the planets of our solar system has an orbital angular momentum that points in identically the same direction as the others (perpendicular to the orbital plane, known as the "plane of the ecliptic," and defined by, you guessed it, the right-hand rule). Most of the angular momentum of the solar system is carried by the largest outer planets: Jupiter, Saturn, Uranus, and Neptune. The Sun has the same orientation of its spin angular momentum as the planet's orbital angular momentum. In addition to revolution, nearly all of the planets spin around axes that are oriented more or less in the same direction as their orbital revolution (however, as an exception, Venus counterspins). This is strong evidence that our solar system developed from a common interstellar cloud that was itself circulating dust and Titan debris, carrying the primordial angular momentum that now defines the planetary orbits. This original angular momentum was conserved as the solar system formed and has remained imprinted in the individual angular momenta of the planets and the Sun. The Sun, over its life span, has actually lost a significant amount of its original spin angular momentum due to the emission of cosmic ray solar winds, which have dissipated the solar angular momentum out into space.

Data obtained over the past hundred or so years confirm the law of conservation of angular momentum on the macroscopic scale of galaxies, planets, and people and their machines, as well as on the microscopic scale of atoms and elementary particles. Thanks to Emmy Noether, we learn that these data imply that space is isotropic—there is no preferred direction in space. All directions in space are equivalent, related by rotations, which constitute a symmetry of the laws of physics. The phenomenon of angular momentum is crucial to our understanding of molecules, atoms, nuclei, and the basic building blocks of matter, the elementary particles. Angular momentum ultimately leads to spooky quantum phenomena and bizarre consequences for the behavior of matter in extreme conditions, issues to which we'll return later.

chapter 6

INERTIA

> SALVIATI: . . . Now tell me what would happen to the same mov-
> able body placed upon a surface with no slope upward or down-
> ward.
> SIMPLICIO: Here I must think for a moment about my reply. . . . I
> cannot see any cause for acceleration or deceleration. . . .
> SALVIATI: Then, if such a space were unbounded, the motion on it
> would likewise be boundless? That is perpetual?
> SIMPLICIO: It seems so to me.
>
> —Galileo Galilei,
> *Dialogue Concerning the Two Chief World Systems*

There is a passage from Galileo's *Dialogue Concerning the Two Chief World Systems* in which his protagonist Salviati, a dissident believer in the heretical theory of the Sun-centered solar system of Copernicus, discusses his beliefs with the conservative Simplicio, champion of the Earth-centered universe and the misbegotten laws of motion of Aristotle, the adopted canon of the Catholic Church. It was written in vernacular Italian, selling out its printing before it was banned, and Galileo himself was brought before the Inquisition. This passage was for the common person, however, an entertaining, satirical, and plain explanation of the principle of inertia.[1]

Modern science, indeed our modern world, begins with this principle. It is the most important law of nature known to us. It can be restated in the form Newton used as his first law of physics: *an object at rest or in uniform motion in a straight line will continue at rest or in uniform motion in a straight line unless acted upon by an external force.* This is the most basic statement that we can make about motion. We say that it is the fundamental principle that governs motion.

In fact, the laws of physics that the object experiences are the same, or invariant, in any uniform state of motion; therefore the principle of inertia is really a *symmetry of nature*—the symmetry, or equivalence, of the laws of physics in all states of uniform motion, for an object, for us, for our laboratory, for anything whatsoever. Galileo understood it in this way. He identified the key concept of the *inertial reference frame* as seen through his characters debating the dropping of rocks off the high masts of ships, both at rest and in uniform motion on quiet seas.

As we attempt to explain concepts like inertia and to connect them to the idea of symmetries, we derive insights into the relationships between different things in the physical world, and we encounter a new and expanded vocabulary. We will never, however, really know *why* there is a principle of inertia or *why* any of the related symmetry principles exist in nature at all. The best that science can ever do is to *notice* things—and to notice how they are sewn together and how they are interrelated—and perhaps figure out how to describe and use them. We'll always be left with yet another unanswered *why* as well as many other things yet to be explained. Although we'll likely never really know why there is inertia, we must certainly notice *that* there is inertia.

Richard P. Feynman was one of the greatest theoretical physicists of the twentieth century and remains a hero to many scientists, including the authors of the present book, to this day.[2] As a youngster Feynman was precociously curious about the world and often performed many of his own home-brewed experiments. He later often recalled his fond relationship with his father, who encouraged this exploration of the world in a rather original way. One day, as a very young boy, Feynman noticed inertia. He was lucky that his "Pop" was there to instill a sense of mystery about this little discovery. The following account illustrates all of the elements of healthy science and healthy mentoring at work:

My father taught me to notice things. One day I was playing with an "express wagon," a little wagon with a railing around it. I had a ball in it, and when I pulled the wagon, I noticed something about the way the ball moved. I went to my father and said, "Say, Pop, I noticed something. When I pull the wagon, the ball rolls to the back of the wagon. And when I'm pulling it along and I suddenly stop, the ball rolls to the front of the wagon. Why is that?"

"That nobody knows," he said. "The general principle is that things which are moving tend to keep on moving and things which are standing still tend to stand still, unless you push them hard. This tendency is called inertia but nobody knows why it's true."

Now that's a deep understanding. He didn't just give me the name [inertia]. He went on to say, "If you look at the wagon from the side [when you start] you'll see it's the wagon you're pulling and the ball stands still. As a matter of fact, from the friction, the ball starts to move forward a little bit in relation to the ground . . . [but the ball] doesn't move back."

I ran back to the little wagon and set the ball up again and pulled the wagon. Looking [from the side] I saw that indeed he was right! Relative to the sidewalk the ball moved forward a little bit [as the wagon moved forward a lot, and then the back of the wagon hit the ball].[3]

Feynman's experience illustrates a simple experiment that anyone can do, at home or in a classroom. Indeed, there are many "experiments" that illustrate inertia, since we are always experiencing inertia. When we accelerate in our car, or in an airplane at take-off, we are pushed backward in the seat. We are the physical objects that tend to remain at rest, and we are acted upon by the force exerted upon us by the backs of our seats. When we try to stop quickly, like slamming on the brakes in our car, we, as physical objects, tend to remain in a uniform state of motion, and we tend to fly forward, unless we experience the force of our seatbelts bringing us to rest with the car. Tripping on a wet patch on the floor, or on the edge of a loose carpet, provides a typical example of inertia, as our feet suddenly halt their forward motion while our upper body continues to move forward with inertia.

Nonetheless, the principle of inertia does indeed seem somewhat mysterious. Though inertia really can be "noticed" with just a little effort, it seems to be forever subtle, lurking in the background and almost transparent, unless its consequences involve something instantaneous and dramatic, like a calamitous accident. We've evolved and adapted to ever-present inertia, so we can navigate the physical world without stopping to notice it and without having to adapt continually to its effects.

So, why was it not until quite late in history—toward the end of the

Renaissance, in fact—that human scholars *finally* noticed inertia? There were certainly plenty of smart people around in many cultures and civilizations throughout earlier centuries, including great Greek philosophers from Pythagoras to Archimedes. Yet they were evidently all confused about inertia, the most basic property of motion. What made inertia so hard to understand, even for the greatest thinkers and observers of antiquity?

The ancient Greek philosophers, who invented geometry, sought to explain how everything in the physical world works. In this pursuit they viewed symmetry as a fundamental guiding principle, just as we do today, inherited from their geometric tradition. If a natural phenomenon, such as the motion of the planets in the sky, could be explained by a theory containing symmetry, then the explanation was all the more satisfying. The theory would then have revealed a deep inner truth about nature. The theory itself would become all the more believable.

A BRIEF HISTORY OF INERTIA, SYMMETRY, AND OUR SOLAR SYSTEM

Frictionless motion, and the concept of an ideal vacuum, however, was too great a conceptual leap at the time of the Greek philosophers. Everyday experience was sculpted by sweating, grunting, and groaning while moving heavy stones or olive oil urns, in wooden wagons with worn wheel bearings (there were no hard hats or safety boots, either). Heavy objects certainly do not appear to execute "uniform motion in a straight line" *unless* one applies a force to them. All things in motion tend eventually to a natural state of rest—so said Aristotle. Mass seemed, for the most part, a measure of a thing's tendency to return toward a state of rest and to produce grunting and groaning when lifted, pushed, or pulled. The Greeks lived in a world dominated by friction; inertia was too hard to notice. They could not separate the concept of friction from the concept of pure and distilled, or idealized, motion. This, we believe, is why they got the most basic ideas of motion wrong.[4]

Contrast this to Feynman's experience as a boy, who noticed inertia using his little toy express wagon and a ball. Even this simple experimental apparatus, however, represents a high level of modern technological achievement. A toy wagon may have oiled, steel-pressed, frictionless bearings supporting a steel axle, with precision cast wheels and tires that allow it to be pulled easily. It rests upon the fairly smooth surface of a paved sidewalk, not a hand-built cobblestone street. The ball in the wagon experiment is an inexpensive, readily available perfect sphere, perhaps a tennis ball from the local dime store. All of this is a product of a modern era: widespread commercial technology that is inexpensive and easily available to everyone, including a boy genius who grew up in the middle of the Great Depression, with an insightful, patient, and caring Pop whose son would one day discover quantum electrodynamics. The ancient Greeks simply didn't have this kind of infrastructure.

Yet as the ancient Greeks turned away from the friction-dominated terrestrial world and cast their gaze to the heavens, there appeared to be something quite different. Planets appeared to move through the sky, as do the Sun and Moon and stars, in certain regular patterns. Symmetries of shape, motion, time, and space seemed to be in evidence up there (hence, perhaps the gods themselves?). Evidently something divine, or of divine intent, was *pushing* the planets along their trajectories. So, in attempting to explain the universe, the thinkers of antiquity invoked a kind of divinity—*symmetry*—as the defining principle for the pushing of planets through the heavens. This culminated in Plato's, and eventually Aristotle's, elevating the idea of perfect circular motion to the compulsory defining symmetry principle of astronomy.

From Pythagoras, born about 569 BCE, to the time of Aristotle, born in 384 BCE, geometry and reason were honed as tools to understand natural phenomena. As we have noted, the solar system's configuration was correctly understood by the astronomer Aristarchus. He had the planets and their orbits in the right places, the Moon orbiting the Earth, and the Sun at the center of the whole system. Aristarchus was the ancient forerunner of Copernicus.

But what followed became for various reasons in large part antiscientific mumbo-jumbo, leading to dogma. The Greek golden age began to crumble, subject to political and economic upheavals. Plato and Aristotle were largely cynical of rational mathematical astronomy and science, preferring instead a kind of faith-based natural philosophy, and they seemed to advocate the benefits of a perfectly ordered society with authoritarian rule. Any such interpretations of these men were amplified by the subsequent rise of an extremist, conservative, powerful, and doctrinaire Neoplatonistic school. "Physics is separated from mathematics and made into a department of theology," said twentieth-century historian Arthur Koestler.[5] Aristotle resolutely imagined an Earth-centered universe and the divinely perfect symmetry of the circular orbit as the governing principle of the heavens. He eulogized the circle and the sphere as the perfection of symmetry and proclaimed that all astronomical objects—Sun, Moon, planets, stars—were perfect spheres. This ultimately became imbued in the canon of an authoritarian Catholic Church. Many subsequent scholars then attempted to reconcile this structural outline with the observed motions of the planets rather than question it.

In the second century CE there lived a Greek astronomer in Alexandria, Egypt, named Claudius Ptolemy. Ptolemy assimilated the Aristotelian philosophy and proposed a theory that became the "standard model" of the universe. This was a mathematically precise theory that lasted almost fifteen hundred years (this is certainly a record for all physical theories to date, not counting religions). Ptolemy's theory, following Aristotle, postulated that the Sun, Moon, stars, and planets all go around Earth. He asserted that hell exists, located at Earth's center, and that heaven could be found in the extreme outer fringes of this cosmological system.

Whereas the Sun, Moon, and stars indeed *appear* to go around Earth in daily circular orbits, the planets migrate across the sky, reckoned relative to the stars, but often change their motion relative to the stars, sometimes reversing direction (we say the motion is *retrograde*), then turning around again and proceeding in the normal direction (*prograde* motion). The concept of *epicycles* was thus introduced by Ptolemy to explain this alternating retrograde and prograde behavior of the planetary motion relative to the "fixed stars." Epicycles were actually an idea borrowed from earlier Greek philosopher Hipparchus. Epicycles are circles upon circles to which the planets were imagined to be attached and along which they moved, like little figurines in an elaborate cuckoo clock.

The universe was thus thought by Ptolemy to be something like a gigantic clock, with things being pushed through their orbits, attached to their epicycles, by some grand, hidden clocklike mechanism, the handiwork of the gods. Some major puzzles remained in Ptolemy's theory, however: the brightness of the planet Venus, for example, inexplicably varied as it moved through the sky. Almost everything else, though, could be fit into this description, ultimately involving only circles and epicycles. Remarkably, the Ptolemaic theory, after considerable perfecting and refinement (we call this "fine-tuning"), did make fairly precise predictions for the future positions of all astronomical objects in the sky.[6]

Thus, a connection with the ubiquitous symmetrical object, the essential ingredient of planetary motion as ordained by Aristotle—the circle—and the precisely measured motions of the planets, was established by

Ptolemy's theory. Here was a cosmology, a "scientific" view of the cosmos, fundamentally based upon the concept of circular, albeit epicyclical, symmetry. Ptolemy's theory was actually useful, as good theories are. Since it generated precise predictions (the ephemerides) for the positions of planets, stars, the Sun, and the Moon, it was useful for navigation and crop planting—as well as astrology (which had a market value, and still has, but is otherwise of significantly less value than horse manure). The Ptolemaic theory was an aesthetically satisfying description of the cosmos, in harmony with the philosophies of Aristotle, and was embraced by the omnipotent Catholic Church. It featured a divine symmetry expressing itself through the circle, revealed to us directly in the motions of the orbs in the sky.

Nonetheless, the elegant theory of Ptolemy, the "standard model" of the first fifteen hundred years of the Common Era—the record-setting, longest-lasting theory of the Universe—is *dead wrong!*[7]

Nicolaus Copernicus, a Polish theologian, significantly revised the picture of the solar system in 1530 in his work *De Revolutionibus Orbium Coelestium* (On the Revolutions of the Celestial Spheres). He rediscovered the lost solar-system configuration of Aristarchus, formulated nearly two millennia earlier, and proposed that Earth *rotates on its axis*, giving rise to the *appearance* that the Sun, stars, and other planets orbit our own planet. The Sun was at the center of everything, and all planets, including Earth, moved about the Sun. The Moon was special and indeed orbited the Earth, and the stars were "fixed" at great distances away from this "solar system." The prograde and retrograde motion of the planets was now a consequence of the fact that Earth is also in an orbit around the Sun, so our vantage point *relative* to the planets is always changing. Hence, Copernicus's theory did away with the epicycles as a fundamental ingredient of the theory. The theory is clever and subtle—the prograde and retrograde motion of planets had become an *apparent effect*, not a fundamental one.

Copernicus never really championed his own theory. He may have been intimidated by the Catholic Church and didn't publish his theory until the very end of his life. The book flew right in the face of biblical scripture, particularly the book of Joshua, where the Lord causes the Sun to "stand still" in the middle of the sky for an entire day.[8] A remarkable disclaimer on the back of the title page of *De Revolutionibus* proclaims, "These ideas are merely hypothetical constructions for predicting the positions of planets and should not be assumed to be true or even probable." This is believed to have been added anonymously by a contemporary of Copernicus, theologian and proofreader Andreas Osiander.[9] Osiander may have done this to protect scholars who could then possess the book without fear of committing heresies. Needless to say, the theory really stirred things up.

The orbits of all planets in the Copernican theory were initially assumed to be circular, preserving the key element of symmetry in Aristotelian philosophy. Now one could, however, understand the changing brightness of Venus, again an apparent effect, associated with its relative orbital position to that of Earth, sometimes close to Earth and on the near side of the Sun, sometimes far away and on the far side of the Sun. The theory also explained the fact, as Galileo later observed through his telescope, that Venus has phases like the Moon, which is completely inexplicable in Ptolemy's theory. The Copernican theory gave a conceptually cleaner and more aesthetic view of orbital motion. The universe of Copernicus is a tricky and subtle place: because the apparent positions of things are determined by the position of the observer, everything cleverly sorted itself out. This neatly explained the apparent prograde-retrograde motion of the planets. Copernicus had banished the epicycles.

By our modern standards of "naturalness," one might think that anyone in his right mind would immediately discard the more contrived theory of Ptolemy, with its epicycles, in favor of Copernicus's sensible and economical description of the universe. But that didn't happen for some time. It is easy, with our 20/20 hindsight, in the era of space travel, to find the whole Ptolemy theory and the behavior of its various protagonists in this historical episode to be laughable. To be objective, however, the Copernican model with its circular orbits was initially *much less precise* than Ptolemy's theory in predicting the positions of the planets in the sky at future dates (we thus say that Ptolemy had a "better fit to the data"; see note 6)! The Copernican model required still further refinements.

The correct scientific description of the configuration of the solar system may have been contained in the theory of Copernicus, but how could one prove that it was right? The publishers of the ephemerides of the day still used the more precise theory of Ptolemy; hence, judged on the objective merits of its accuracy, Ptolemy's theory won. Moreover, the religious leaders rejected this non–Earth-centered, non-Aristotelian, Copernican theory altogether, ultimately deeming the teaching of it to be a heresy, punishable by death. The Copernican theory involves subtleties of perspective that fool the observer into "seeing" the epicycles that aren't really there—this may have seemed unsettling to religious leaders concerned with the role of the devil in contaminating the purity of faith. Ultimately, they wanted to maintain their central ruling authority and repel any assault on their chief mentor, Aristotle, and the divine symmetry of the circle.

Here we encounter Giordano Bruno. Bruno was smitten with the self-evident rational elegance and logical beauty of the Copernican theory, which so democratically did away with the privileged position of Earth at the center of the universe. He widely and vociferously proclaimed that the entire solar system itself was just one of many such solar systems in a greater universe. Bruno's universe was thus filled with many other similar systems, spanning a virtually infinite void. He further proposed that there likely would be other inhabited worlds with intelligent beings equal, or possibly superior, to ourselves, living out in the vast universe. Bruno was, in a sense, the first modern cosmologist, who had foreseen the enormous *homogeneous* and *isotropic* universe of our modern cosmology—that there is, in reality, no center or preferred direction in the vastness of the universe. For these outrageous blasphemies, Bruno, as well as other dissenters, was tried in the Inquisition and ultimately burned at the stake in 1600.

Johannes Kepler then emerges in the story. Kepler had the firm conviction that the Copernican theory must be the true configuration of the solar system. At around the same time Bruno paid for his beliefs with his life, Kepler tackled the problem of reconciling the difficulties that the Copernican theory had with the data on planetary motion. He believed that if he could find out why the Copernican model gave incorrect predictions, and if he could fix it, he would discover a magnificent new symmetrical structure to the universe. Kepler therefore had some strong preconceptions, but he proved to be intellectually honest. He had access to the most precise astronomical measurements of the era, due to his association with difficult but gregarious astronomer Tycho Brahe, for whom Kepler had been a research assistant. What we owe to Kepler, in particular, was his great scientific integrity and persistence, for he was the true champion of scientific truth. He demanded a strictly correct explanation of planetary motion, one that fit the data precisely, based upon the Copernican theory, whether it ultimately met with his philosophical inclinations or not.

Kepler first noticed that the geometrical center of Earth's orbit was not the Sun but rather a point in space some distance from the Sun. He then focused most of his attention upon the confusing motion of Mars.

Kepler first confirmed that its motion lay in a plane, though inclined by about two degrees from the plane of motion of Earth. The motion of Mars was seen through the detailed and precise measurements of Tycho Brahe to be clearly departing from a Sun-centered circular orbit. Kepler, however, discovered the correct geometric form of the orbit to be not a circle but an ellipse, ultimately proving that all of the planetary motions in Copernican theory are such ellipses. Finally, the speed of the planet during its orbital motion was not constant but was rather seen to vary—another Aristotelian precept proven wrong. Kepler discovered the correct relationship between the planet's speed and position of the planet in the orbit. These discoveries, a tour de force of logic and research, were observational and indisputable facts now implied by the Copernican solar system, though it was not a satisfying outcome for Kepler, who was seeking deeper divine symmetries and Pythagorean mathematical perfection in the laws of motion of the planets. The facts, however, were in: *we must give up the symmetry of the circle to have the planets orbit the Sun in the solar system.*

It should be noted that, at this time, Ptolemy's theory was completely inconsistent with Tycho's precise observational data. Kepler ultimately deduced his set of three laws that completely defined the motion of planets in their orbits, publishing in 1609 and 1619 the sequence of considerations that led to his discoveries. For one, he concluded that a planetary orbit, not a circle but actually an ellipse as mentioned above, has the Sun at a single focus of the ellipse (see fig. 11). Kepler also deduced a precise mathematical law governing the length of time of orbital motion through any part of the orbit—this "second law of Kepler" actually represents the discovery of the conservation of angular momentum (which we encountered in the previous chapter). And finally, he found that the period of the orbit, T, is related to the size of the orbit, R, through the mathematical relationship "T^2 is proportional to R^3," with the same constant of proportionality for all the planetary orbits.[10] The complete and detailed schematic of the motion of planets was now known. The Keplerian modifications and specific properties of the orbits in Copernican theory now gave perfect agreement with the most precise astronomical observations and was now perfectly predictive, whereas Ptolemy's theory was not. The publishers of the ephemerides could now use the Copernican-Keplerian solar system with greater reliability than the scheme of Ptolemy. From a scientific vantage point, Ptolemy's theory was now dead.

An ellipse is a well-defined mathematical shape, forming a kind of

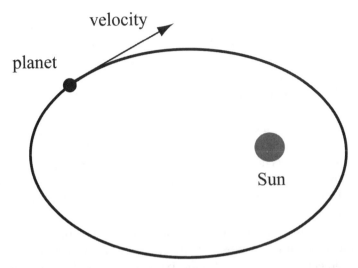

Figure 11. Elliptical planetary orbit, with the Sun at one focus of the ellipse; the planet's instantaneous velocity is tangential to the ellipse.

"squashed," or less perfect, circle. One evidently needed to abandon Aristotle's basic driving notion of symmetry, as expressed through the circle, to correctly explain the universe. Kepler had uncovered a complete and correct set of rules that truly described planetary motion. But what had become of symmetry? Symmetry now appeared marginalized, or at best, only approximate, the perfect circle becoming squashed into the imperfect ellipse. Kepler's laws, however, were exact and complete, setting the stage for the next set of questions. Lurking within Kepler's laws of motion was a new form of symmetry residing at a much more profound level within nature.

We note that Kepler's description of the solar system was a *phenomenological* theory. Phenomenological theories are commonplace in science. They are accurate sets of rules that describe a specific phenomenon, or object of study, but they often make no deeper connections to the rest of science. Nevertheless, they can aid scientific progress because they reduce data obtained through many observations to an economical set of only a few rules. The next step needs only to explain this set of phenomenological rules. However, the acceptance of Kepler's theory would be problematic in his day, dividing down political and religious lines, clashing with the laws of the church. As noted, the Catholic Church held that any view contrary to the ancient Ptolemaic theory was a heresy, punishable by the worst forms of torture and/or execution. And they meant it—the fate of poor Bruno, and others, was very much on the minds of the practicing scientists of the early seventeenth century.

NOTICING INERTIA

Overarching the Copernican theory of the solar system, as refined by Kepler, there remained a deeper and seemingly inaccessible scientific question: what makes the planets move in their orbits? The force acting on a planet would seem to have to be *pushing* in the direction of motion of the planet. We say that this direction is *tangential to the orbit*. This was the idea inherited from the Greeks and their friction-dominated experience of grunting and groaning while pushing carts of olive oil urns. Without something "pushing" the planets they would obviously come to rest, as do carts drawn by mules, or rocks being moved to a building site, and so on. Things have a "natural tendency" to come to rest—Aristotle had said so. Kepler had no better answer to the question of what makes the planets move. It is often said that his reply was that "angels are beating their wings and pushing them"; however, Kepler actually evolved a complicated and contrived theory of vortices emanating from the Sun that both pushed the planets along like a broom and yet held them in their elliptical orbits.[11]

The key to the future of science turned on this question. And, it was Galileo who finally freed human thought from the friction-dominated world of the Greek philosophers. Galileo was a remarkable scientist, possibly the greatest who ever lived. He discovered many things that radically changed our view of the universe. He was, after all, the first to observe the sky with a 20× telescope, which he had methodically constructed in his own laboratory, in 1609 at the University of Padua in the Venetian Republic, now Italy. He was both an extremely skilled experimentalist—who built precision scientific instruments and made detailed and painstaking observations with them—and a theorist who would reason the connection of his observations to deeper universal principles. Galileo discovered that the Moon had mountains and craters, that Jupiter had moons, that the Sun rotated on its axis and had spots, that Saturn had rings, and that Venus was enshrouded in clouds and displayed phases like the Moon, completely as predicted by Copernicus's theory. It is amazing what nature offers up to the first kid on the block with a powerful new scientific instrument!

Nothing, however, is more significant than Galileo's distillation of his many observations about motion. He undertook a series of experiments with objects moving on smooth and frictionless surfaces, with pendulums, and with objects dropped from heights or rolled up and down inclined planes. Galileo, as we have said, was evidently the first person to notice inertia and systematically study it. He isolated and removed friction from his observations of motion and discovered that inertia remained. Galileo distilled his observations of motion and thus discovered the principle of inertia.

From the principle of inertia we realize that *nothing* "pushes" the planets around in their orbits—they are in an eternal state of motion by virtue of inertia. There is *no* friction in the vacuum of space, and it is friction that hides the tendency for something to move eternally with a constant speed in a fixed direction in space. Friction is the very common force that changes the state of motion of a heavy wagon laden with olive oil urns, causing it to come to rest. Remove friction, and the wagon will move uniformly in a straight line, forever.

The principle of inertia predicts, however, that the planets will move in straight lines unless some force alters the direction of motion. In this case, the force is pulling the planets toward the center of the orbit about which they revolve. For example, a stone attached to the end of a string and whirled around in a circle is pulled into the circular motion by the force of the string acting upon the stone. This force acts toward the center of the stone's "orbital motion." If the string breaks, the stone will fly off in a straight line tangent to the original circle, obeying the principle of inertia. And so it is with planets. But what is the force that pulls the planets toward the centers of their orbits? *Force*, by the friction-dominated, urn-carrying experience of the Greeks, was required to produce motion but is really required only to *change* motion—that is, to *stop* or to *start* motion, or to change the state of motion, its direction and speed. Changing motion is called *acceleration*. The force of friction is ever-present to stop motion in our everyday world but not in the vacuum of space, in which the planets move. Friction is thus a consequence of the complexity of the world around us, but once accounted for, the law of inertia holds everywhere around us and at all times throughout the entire universe.

Incidentally, Galileo was brought forth before the Inquisition in 1633. He was threatened with torture and death, shackled, and forced on his knees to renounce his beliefs in the Copernican theory and all of his observations with the telescope, through which his prosecutors refused to have a look for themselves. Galileo was sentenced to imprisonment (actually, house arrest) for the remainder of his days.

We surely hope those times are gone forever, but we're not certain that they are.

THE UNION OF SYMMETRY, INERTIA, AND THE LAWS OF PHYSICS

The principle of inertia is a symmetry. As we have seen, symmetry is an equivalence between things. The principle of inertia refers to an equivalence that is the equivalence of all states of uniform motion. That is, any state of uniform motion of an object stays the same unless something intervenes to change this state of motion—in other words, unless that object is acted upon by a force.

Now, we can restate this in a more profound way. All states of uniform motion are really equivalent to each other, and this is a symmetry of nature! It is called *Galilean invariance* (or *Galilean relativity*, or *Galilean symmetry*): *all states of uniform motion are equivalent for the description of physical phenomena*. What do we mean by the statement "all states of motion are equivalent to each other"? If I am standing still and you are moving, we don't appear to be in equivalent states of motion? This, however, is *not* the meaning of the statement.

What we do mean is that the laws of physics are *exactly the same* as seen by different observers, each in her own state of motion. So, uniform motion is a symmetry of the laws of physics. If an observer in her laboratory is moving uniformly through space, the laws of physics she experiences will be no different than for another observer in his laboratory standing still. In fact, there is no absolute meaning to such concepts as "standing still" and "moving uniformly through space." The observer I see, moving uniformly, sees himself to be standing still, and he sees me to be moving uniformly in the opposite direction. There is no possible way to define which observer is absolutely moving through space, and only their *relative motion* can be determined. We would both find that within our own laboratories the laws of physics are identical. We have a name for uniform states of motion of hypothetical laboratories—we call them *inertial reference frames*.

We can see how the law of inertia is encoded into this symmetry. If something stays at rest in my inertial reference frame, because there is no force acting upon it, then something must also stay at rest in Sue's inertial reference frame, when no force is acting upon it. But Sue is in a state of uniform motion relative to me. Hence, the principle of inertia follows logically: an object in uniform motion (i.e., moving at a constant velocity or at rest, as in Sue's inertial frame) must continue in uniform motion unless acted upon by a force.

You may have noticed the term "relative to" in the preceding paragraph. In fact, the Galilean invariance is actually what is now known as the *principle of relativity*. Later on, when Einstein got into this game, it became more profound. It turned out that we don't even need to insist upon "uniform motion" once we are willing to talk about the laws of physics more generally. This got us to Einstein's general theory of relativity.

Hence, we think of the principle of inertia as a consequence of the equivalence of the laws of physics in all inertial reference frames. In this way, it is a symmetry of the laws of physics—this is its true essence.

NEWTON'S LAWS OF MOTION

As we have seen, before Galileo, people thought that a force *produces* motion—if no force, then no motion. However, we have seen that this is wrong. Inertial motion, that is, motion at a constant velocity, also involves the lack of any forces acting upon the body, just like sitting at rest does. "Forces" are evidently needed to change a state of motion of an object. But what is a force and what, exactly, does a force do?

It was many years after Galileo that Isaac Newton stated precisely what a force is: force equals mass times acceleration. Or, written as one of the most resoundingly famous equations of all time, $\vec{F} = m\vec{a}$. This does not state that a force produces motion, since motion is velocity, as the Greeks evidently conceived it. Rather, a force produces *acceleration*, which is the *rate of change of velocity* (per unit time). Acceleration is a continuous rate of change from one inertial reference frame to another. Note that acceleration, the time rate of change of velocity, must be a vector quantity with both a magnitude and a direction in space. Therefore Newton's law of motion requires that force, $\vec{F}$, also be a vector, with a direction in space and a magnitude.

Newton formulated the laws of nature that define classical physics, stating first the three fundamental laws of motion and force:

1. An object in uniform motion or at rest remains in a state of uniform motion or rest unless acted upon by a force.

2. A force, $\vec{F}$, acting upon an object of mass, m, produces a resulting acceleration, $\vec{a}$, determined by the equation $\vec{F} = m\vec{a}$.

3. If object B exerts a force upon object A, of $\vec{F}_{AB}$, then object A exerts a force upon B, of $\vec{F}_{BA} = -\vec{F}_{AB}$ (that is, $\vec{F}_{BA}$ points in the opposite direction as $\vec{F}_{AB}$ but has the same magnitude; it is called the "reactive" force).

The first law is a just a restatement of the principle of inertia. The principle of inertia evolved in its significance over the intervening years between Galileo and Newton. We're not completely sure who first cast it precisely into this form—it may have been either Galileo (Italian), Newton (English), René Descartes (French), and/or others along the way.[12] By the time of Newton, however, we see that a complete understanding of the interrelationship among inertia, motion, and forces has been perfected, to a fundamental level that describes all phenomena accessible to experiments of the time. This is a grand reduction of all phenomena to a few simple so-called laws of physics.

Newton's second law is often called the "equation of motion." From Newton's second law of motion, if we are given the mass of a particle and we are also told the force acting upon it, we can compute the change in uniform motion (or rest) and determine the subsequent motion of the particle exactly! This is the real power of physics—the ability to predict with certainty the outcome of events. There are many forces with well-defined formulae in physics, and the motion they produce is determined, through this single law, for all of them.

Newton's third law is actually a deep consequence of the translational invariance of the laws of physics, and it leads to the law of conservation of momentum, as we described earlier. It also makes direct contact with Noether's theorem, as we will see subsequently.

ACCELERATION

Let us consider acceleration in slightly more detail. This concept, like the concept of power, is also a little confusing to many people. We can talk about acceleration without mentioning force. Acceleration, put simply, is the *change of velocity with time*. Of course, velocity is the time rate of change of distance traveled. So, acceleration is the time rate of change of the time rate of change of distance traveled. Newton's law of motion equates the direction and magnitude of a force vector to the acceleration that the object will experience. If the force on the object is zero, $\vec{F} = 0$, then the acceleration will be zero, $\vec{a} = 0$, and the object will not accelerate. This means it must have a velocity that does not change in time—a constant velocity, $\vec{v}$; we define this as inertial motion.

Earlier we discussed an automobile out on the open highway traveling at a speed of 30 meters per second (abbreviated m/s; equivalent to 60 miles per hour). We took our foot off the accelerator and let the car coast. We measured how long, in seconds, it took for the car to slow down to 25 m/s (50 mph). We found that it took about 10 seconds for this change in velocity to occur. During this time the car was accelerating, but of course, the acceleration was directed in the opposite direction to the car's velocity—the car was *decelerating*, or slowing down (deceleration being just negative acceleration). This acceleration is just the velocity *after* (25 m/s), minus the velocity *before* (30 m/s), divided by the time interval (10 seconds), or $(25 - 30)/10 = -0.5$ m/s^2, which shows that the unit of acceleration is a *length scale divided by the square of a timescale*. Like velocity, acceleration must also be a vector, which in the present example either points in the direction of the velocity if it is positive (when we are speeding up) or in the opposite direction (when we are slowing down, or decelerating).

We can also do another experiment, one that has a great deal of appeal to teenage boys: we can measure how fast we can get our car from rest to 60 mph (30 m/s). This should be done carefully in a safe, open-road environment, by a very experienced driver. One simply puts the "pedal to the metal" and measures in how many seconds the car can reach the velocity of 60 miles per hour. For a typical four-cylinder compact car, we find that it takes about 8 seconds, so we find that the car is accelerating at about 3.8 m/s^2 (i.e., 30 meters per second, divided by 8 seconds).

Now, suppose we drop an object and watch it fall. We'll see that the object accelerates toward the ground at 1 "g," which is a rate of acceleration of about 10 m/s^2. (It is fun to devise a simple experiment to measure g, and many examples can be found on the Internet.) So, the automobile in our example can accelerate at approximately 38 percent of g. Many automobiles can accelerate much faster than this (such as police cars). Still, this is "comfortable acceleration" and does not cause serious side effects after a prolonged time for most people (unless they hit another vehicle).

Now, here's an intriguing question. Suppose we built a supertrain that accelerates continuously at the reasonably comfortable rate of, let's say,

0.5 g, or 5 m/s^2, after departing Chicago, halfway to New York City. Then, at the Ohio-Pennsylvania border, the train reverses its acceleration, decelerates, and finally comes to rest in New York City. How long would this trip take? The answer: about 16.3 minutes![13] A businessman in Chicago could schedule an impromptu meeting with the New York City office in an hour. Without even having to pack clean underwear and a toothbrush, he could go down to the "LaSalle Street Superstation," flash a credit card, and hop on a single supertrain car, shaped much like the fuselage of a small jet aircraft, like a Boeing 737, with a seating capacity of about one hundred people. Every ten minutes, as the car would fill up, the doors would automatically close, and the train would pass through an airlock into a tunnel that had a 0.01-atmosphere vacuum. Then it would comfortably accelerate, levitated on a system of superconducting magnets, driven forward by magnetic induction. After 490 seconds, about 8 minutes later, the train would be deep under the Ohio-Pennsylvania border, traveling in the vacuum tunnel at a speed of $v = at$, or 5,400 mph! Then it would begin to decelerate gently, and about 8 minutes later it would roll to a stop in the brand-new underground terminal in the Lower Manhattan business district. All of this is, in fact, possible with 1950s technology—and a fraction of the 2005 US military budget.

In these examples the acceleration, hence the force, has been aligned with the direction of the velocity of the particle (the "particle" is an automobile or a Maglev train). However, in another example, a force can act perpendicular to the direction of the velocity as well, causing the acceleration to be perpendicular to the velocity by Newton's law. In this case, we get a deflection of the motion from a straight line. If the acceleration is always constant in magnitude and always perpendicular to the velocity, we end up with motion in a circle.

Therefore the planetary motion in approximate circles in the Copernican theory means that the force is perpendicular to the velocity of the planet. Certainly angels are not "pushing" the planets along in their orbits after all. Rather, the planets are moving because something in the remote past started them off (the supernova blast of a Titan, which set the original planetary debris cloud in swirling motion), and inertia keeps them going. The force is tugging them away from motion in straight lines. The force vector is pointing toward the center of the orbit. If we look at the center of the orbit, we exclaim, "Eureka!" It turns out that the force vector is pointing toward the Sun! The force causing the planets to accelerate in their orbits is therefore produced by the Sun (see fig. 12).

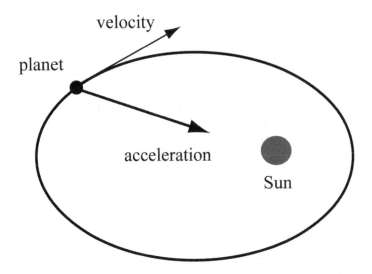

Figure 12. For an elliptical planetary orbit, whose motion is determined by Kepler's laws, Newton found that the planet's acceleration vector points directly at the Sun.

The achievements of Copernicus, Kepler, Galileo, Newton, and others marked a spectacular epoch in human history. The time of Newton is often called the beginning of the *Enlightenment* (usually considered to be the eighteenth century) because of the profound changes in political philosophy, discovery and trade, technology, our understanding of the world's geography, and especially the new and correct understanding of physical motion and physical forces, and the scientific method and reason that led to it all. The elucidation of the classical laws of physics ultimately brought forth a period in which the "first world" industrialized, resulting in an unprecedented prosperity coming to the common people, together with political rights and new standards of government. It ultimately led to the steam age, to steel manufacture, to electric generators, to electric motors, to telegraphy, to radio, and to electric illumination, to name a few. The Enlightenment was a period in which purely academic and esoteric research activities translated into the empowerment of the human species.

GRAVITY

Yet we are not finished with our discussion of force. We have seen that there is a force exerted on the planets by the Sun that holds them in their circular orbits around it. What is this ubiquitous force? It is called *gravity*.

With Newton's precise law of motion we can ask the following scientific questions: What is the nature of the gravitational force between planets and the Sun that deflects the motion of the planets from a straight line into their elliptical orbits? Why ellipses? What is the precise mathematical form of the gravitational force?

Newton solved these problems. By using Kepler's laws of planetary motion, Newton determined that the acceleration vector of the planets always points directly toward the Sun (with negligibly small corrections due to the other planets, such as Jupiter, Saturn, etc.). The magnitude of the acceleration of a planet was found to be *inversely proportional to the square of the distance of the planet from the Sun*. The magnitude of the acceleration had nothing to do with the mass of the planet under consideration! Hence, he thought, the force holding the solar system together must be due to the Sun itself, pulling on the planets to deflect them from otherwise moving, according to inertia, in a straight line. Newton's great realization was that Earth similarly exerts a weaker force upon the Moon, pulling it toward Earth and deflecting its inertial motion into a closed orbit. And, finally, Newton realized that this same force holds all things to the surface of Earth—rocks, water, air, people—as it pulls toward Earth's center. This explains, for example, why an apple falls from a tree toward the ground. It is profound and remarkable that the force acting throughout the solar system, creating planetary orbits, is one and the same as that found here on Earth, shaping mountains, seas, grasses, and trees. Newton was led to the *universal law of gravitation*.

Let's dissect Newton's universal law of gravitation. This will be an exercise in reading—reading a math formula. This is easier than learning to read French (which is not so hard, either) and requires only a little patience.

According to Newton, the *magnitude* of the force of gravity exerted upon object A by object B is called F_{AB} and is given by the formula

$$F_{AB} = \frac{G_N m_A m_B}{R^2},$$

where R is the separation between them. Often such formulas produce a malfunction in the reader called "eye glaze"; please blink twice and continue reading . . . it will all make perfect sense in a moment.

The universal law of gravitation is an example of what is known in physics as an *inverse-square law force*, that is, a force that falls off in magnitude, or strength, with distance, like $1/R^2$. The electric force between two stationary electric charges is also an inverse-square law force.

The force is a vector and must therefore also have a direction. We could write a better formula that illustrates that, but words suffice. Object A experiences the force of gravity, with the magnitude we have written, but the force points as a vector at the direction of object B. And, by symmetry, object B experiences the same magnitude of force, which points in exactly the opposite direction, back to object A.

In this formula m_A is the mass of object A, and m_B is the mass of object B. This means that the force of gravity is stronger between two very massive objects than between two very low-mass objects. For example, if A is Earth, we substitute $m_A = m_{Earth}$, and if B is the Sun, we substitute $m_B = m_{Sun}$ into the formula. Thus, if we could somehow double the mass of the Sun, holding everything else fixed, then the force of gravity that Earth would experience from the Sun would become doubled, and Earth's orbit would change, becoming a "tighter" ellipse, with a smaller average distance from the Sun.

Notice that the formula is completely symmetrical between body A and body B. That is, if we swap A for B everywhere, we get the same result for the magnitude of the force between the two objects (and the directions are correspondingly swapped). All objects gravitate in the same way and feel gravity in the same way. This is why it is called the "universal" law of gravitation.

The quantity G_N in the numerator of the formula is a *fundamental constant*. Newton had to introduce this factor in order to specify the *strength* of the gravitational force. We call this Newton's gravitational constant, or just Newton's constant, for short. The history of the experimental measurement of G_N is quite interesting, but for now, let's just quote the best determined value. This "magic number" of gravity is measured from experiment and takes the value $G_N = 6.673 \times 10^{-11}$ m^3/kg s^2.

Note that G_N is not just a pure mathematical number, like 3.1415, but rather is a *physical number*, because it must be given in reference to a system of units, and its value will be different in other systems of units. We have quoted G_N in the meter-kilogram-second system of units. Indeed, we can write, in nonscientific notation, $G_N = 0.00000000006673$ m^3/kg s^2, and we see that G_N is a seemingly very small number. Gravity, despite its ubiquitous character in nature, is actually a very feeble force![14]

When we are standing on the surface of a large spherical body, usually Earth, with a mass of m_{Earth}, then we are feeling the gravitational pull of all the matter beneath our feet. To compute the force of gravity exerted by Earth on things at the surface, we use for R the distance to Earth's center, that is, the radius of Earth, R_{Earth}. This is true even though the mass of Earth is not entirely concentrated at the center. (This was actually quite a difficult thing to prove mathematically in Newton's era and may have inspired him to invent the integral calculus, which is used nowadays in the most common proofs found in college textbooks.)

Thus, let us consider the acceleration of an apple (viewed as object A) toward Earth (viewed as object B). Putting these two laws together, we have for the magnitude of the force experienced *by* the apple *due to* Earth $F_{apple} = G_N m_{apple} m_{Earth}/(R_{Earth})^2$. This is a vector, pulling the apple toward Earth's center. On the other hand, Newton's second law, the equation of motion, says that the force produces an acceleration of the apple, $F_{apple} = m_{apple} a_{apple}$, therefore $m_{apple} a_{apple} = G_N m_{apple} m_{Earth}/(R_{Earth})^2$. So long as the apple is attached to the tree by its stem, there is a force acting by the tree upon the apple that exactly balances the force of gravity—the apple doesn't move. If the apple is released by the stem breaking, the only force acting on the apple is gravity, so the apple accelerates downward toward Earth.

Now, we learn something remarkable from this apple-Earth problem. Let us compute the acceleration that the apple actually experiences due to gravity. We do this by simply dividing both sides of our above equation by the mass of the apple. We get $a_{apple} = G_N m_{Earth}/(R_{Earth})^2 = g$. This formula says that the acceleration of the apple toward Earth does *not* depend upon the mass of the apple! In fact, the acceleration experienced by the apple is the same for all objects near the surface of Earth; this acceleration is invariant to the size, mass, and shape of the object. The magnitude of the acceleration is something we first encountered with the Acme Energy Company—g, which is our standard symbol for the acceleration experienced by all objects at the surface of Earth due to gravity. We could put the numbers into the formula, the mass of Earth, the radius of Earth (from Eratosthenes' famous result), and the value of Newton's gravitational constant, and we would get the approximate value $g = 10$ m/s^2. Of course, by independently measuring Newton's constant in the lab, the radius of Earth, and g, we can determine the mass of Earth, which is how it has actually been done.

All objects will fall with the same acceleration, g, if you can neglect air resistance. This is, at first sight, an astonishing fact. This also dramatically contradicts an ancient claim of Aristotle, namely, that a ten-pound weight would fall ten times faster than a one-pound weight. The fact that the acceleration is independent of the weight was supposedly demonstrated publicly by Galileo when he dropped two different weights off the Leaning Tower of Pisa. To the best precision of observation available at the time, the two weights supposedly fell to the ground and arrived at the same instant. We actually don't know if Galileo ever did the experiment, but many people to this day intuitively believe that heavy objects do fall faster than light ones.

A standard physics classroom demonstration compares the rate of fall of a penny and a feather, starting from the top of a long glass tube from which the air can be pumped out. Before pumping, the penny drops in less than a second, whereas the feather floats down in ten seconds. After pumping out the air, the race is repeated, and both the penny and the feather fall to the bottom simultaneously. *Apollo 15* astronaut David Scott performed the experiment on the Moon, which has no atmosphere, dropping a feather and a hammer, and both objects fell at the exact same rate (though the acceleration on the surface of the Moon is one-sixth that on Earth, since it involves m_{Moon} and R_{Moon}, plugged into the same formula).

According to symmetry, Earth also accelerates toward the apple, but the magnitude of this acceleration is much smaller than the apple's acceleration, g, by the tiny factor of $m_{\text{apple}}/m_{\text{Earth}}$. So, we can ignore the acceleration of Earth toward the apple. However, we note that the *total momentum* of Earth plus apple is conserved, a consequence of Newton's third law. In fact, we see that the Newtonian law of gravitation makes no reference to any particular place in the universe and involves only the relative position (and direction) of the apple and Earth. We can thus use the same formula in such a distant location as the Andromeda galaxy that we use here in our solar system! The formula is translationally invariant, and therefore momentum, according to Noether's theorem, must be conserved!

Using Newton's law of gravity and a little analysis, we encounter the concept of "gravitational potential energy." An object at rest at the top of a tower has a larger gravitational potential energy than when it is at the bottom of the tower. This means that when the object is initially at rest at the top of the tower, it has no energy of motion (kinetic energy) but a hefty amount of potential energy. The total energy of the object is the sum of these. When it is dropped, the potential energy is reduced but converted to increasing kinetic energy as the object accelerates downward. Nonetheless, the total energy, potential plus kinetic, is always exactly the same—it is conserved.

Newton discovered that the orbital motion of the planets, predicted by his mathematical laws, are indeed elliptical. He completely explained Kepler's phenomenological laws of motion with his universal theory of gravity and the deeper laws of classical physics. This constituted a mathematical tour de force for Newton, requiring that he invent a new system of mathematics, the calculus. In addition to the closed, elliptical orbits of planets, Newton discovered in the mathematics that there are also *open hyperbolic and parabolic* trajectories, corresponding to massive bodies coming in from an infinite distance and being deflected, or "scattered," by the Sun (such as comets).[15] For the solar system there are also detailed corrections to pure elliptical motion due to the gravitational interactions among all the planets. One can also discover new planets, such as the recently discovered planetoid Sedna, beyond the orbit of Pluto, through careful analysis of the wobbles in the planetary orbits with precise data, using Newton's laws.[16] The US space program (the National Aeronautics and Space Administration, abbreviated NASA) in 1969 successfully landed men on the Moon, navigating exclusively by employing Newton's laws of motion.

Ultimately, however, even Newton's theory eventually breaks down and cannot describe phenomena that involve motion approaching the speed of light. In fact, the correct theory of gravity is a stunning and radical revision of Newton's view. Einstein replaced Newton's theory with general relativity, explaining why gravity is universal and how it intimately involves the geometry of space and time. Newtonian physics, however—within its domain of validity—is the correct description of nature, and it is here to stay.

The domain of validity of Newtonian physics encompasses our everyday lives. But it breaks down for objects that are very tiny or that move at speeds approaching that of light. And what replaces it? Bigger and better symmetries, of course.

chapter 7

RELATIVITY

Henceforth, space by itself, and time by itself, are doomed to fade away into mere shadows, and only a kind of union of the two will preserve an independent reality.
— Hermann Minkowski, *Space and Time*

THE SPEED OF LIGHT

Many early philosophers and scientists, such as Aristotle and Descartes, thought that the speed of light was infinite; hence, light would be transmitted instantaneously through space.

Galileo, however, considered the possibility that light traveled with a finite velocity, and he devised a primitive way to attempt to measure it. He would flash a light signal at a very distant observer, his assistant, who would quickly try to flash a second light signal back. This involved ultra-fast reflexes on the part of the assistant, to minimize his reaction time upon seeing the first flash of light and immediately transmit the return signal. Galileo attempted to determine if there was a perceptible time interval between the initial and returning flashes that became longer as the distance between the two increased—this would indicate a lag time proportional to distance, and therefore due to a finite velocity of light. He failed to detect an effect, because human reaction time is much too slow in comparison to the travel time of a light flash over terrestrial distances. Galileo was, nonetheless, able to demonstrate that the speed of light must exceed about six thousand miles per hour (the speed of light actually exceeds this by a factor of about a hundred thousand).[1]

The first detection of the finite speed of light arose in astronomy. The science of astronomy was of paramount importance in the era of the great sea empires and had taken on an official and stately function in France and Britain. In particular, astronomy was essential for global sea navigation and time keeping. Knowledge of latitude and longitude on the oceans was the key to navigation, and, indeed, to survival. Using a sextant, a navigator could fairly easily determine his latitude by measuring the angular height of the Sun above the horizon at the culmination of its trajectory during the course of the day. This point, known as "local noon," is when the Sun is highest in the sky.

Determining longitude, on the other hand, was harder, as it is essentially a time measurement. To determine longitude one needed to know the exact time in Greenwich when observing local noon at sea. For example, if I know it is exactly 1 PM in Greenwich when I observe the Sun to be in my local noon position, then I infer that my longitude is 15 degrees west of Greenwich (one hour corresponds to 15 degrees, because 24 hours $\times$ 15° = 360°, or one full rotation of Earth). Unfortunately, resilient and reliable mechanical seafaring clocks did not become available until much later.[2]

One particularly egregious error in the reckoning of longitude by a British admiral in 1707 cost the navy four warships and two thousand lives when the ships ran aground (not to mention the life of a poor shipmate who had kept the correct longitude, as a hobby, but was hung by the neck at the yardarms for mutiny when he questioned the admiral's reckoning).[3] Many scientists thought the problem of engineering seafaring clocks was intractable and urged the use of "astronomical clocks," that is, any natural phenomena that occurred in the nighttime sky at a regular and predictable time. These would be observable anywhere on Earth, including at sea, and could thus provide a means of absolute time keeping, despite the fact that precise measurement of time in this way was often difficult, requiring perfect weather and painstaking measurements on decks of heaving ships.

In 1676, a Danish astronomer at the Paris Observatory named Ole Rømer was studying in detail the motion of the moons of Jupiter. The most conspicuous satellites of the great planet, later known as the "Galilean moons"—Io, Europa, Callisto, and Ganymede—were discovered by Galileo on January 7, 1610, through his homemade, 20×-magnification telescope.[4] The orbital period of a Galilean moon is like a regular clock pendulum, observable in principle anywhere on Earth with a telescope and clear weather, when Jupiter is visible. This had the potential of providing an unchanging and worldwide standard of time.

Io, the third-largest of the Galilean moons, with an orbital period of about 1.8 Earth days, was a suitable candidate for such a universal clock. Io was eclipsed at regular intervals as it moved behind the planet Jupiter. Io moved in a nearly perfect circular orbit, and it therefore provided a perfect "tick-tock" interval, the "tick" being the moment it disappeared behind the disk of Jupiter, and the "tock" being the moment it reappeared. Rømer discovered a subtle effect, however. He first measured, using his Earth-bound laboratory clock, the expected time of the eclipses—the ticks and tocks—when Earth was at closest approach to Jupiter in its orbit. As Earth swung through its orbit, farther away from Jupiter, he observed that the actual time of the tick-tock eclipses lagged behind their expected time. About six months later, when Earth was at its farthest distance from Jupiter, the eclipses were running a whopping sixteen minutes behind schedule. Then, as Earth came back around in its orbit and again approached Jupiter six months later, the time lag disappeared, and the tick-tock of the eclipses occurred again at precisely the expected time. This cycle repeated throughout the Earth year.

Rømer's discovery was one of those delightful events in the history of science in which the observer, while attempting to measure something mundane in great detail, was treated to a grand surprise. Rømer realized that this annual cycle of "tick-tock" lagging corresponded to the changing distance between Earth and the Jupiter-Io system in its orbit. Indeed, Rømer realized that the correct explanation of the effect is that the light from Io travels at a *finite velocity*—light must traverse a greater distance when Earth is farthest from Io than when it is closest—hence the time lag when Earth is farther away. Rømer had measured a time lag of sixteen minutes; thus light takes sixteen minutes to traverse the diameter of Earth's orbit, and therefore eight minutes to travel the *radius of the orbit*, from the Sun to Earth. To determine the speed of light, c, one needs therefore to determine the distance between

Earth and the Sun (the radius of Earth's orbit) and divide this distance by the eight-minute time interval.

The distance between Earth and the Sun is called the *astronomical unit*, or AU, and it is the most important distance scale in the history of astronomy. The AU establishes the baseline of the triangle used for determining the distances to all of the nearest stars. That is, the AU is the basic "survey instrument scale" of astronomy. Unfortunately, however, the AU is very difficult to determine. The Greeks had tried various ingenious methods but never really obtained any precision, rather only guesses that were off by factors of ten or more.

When we attempt to measure an astronomical distance of something that is not too far away, let's say a nearby star (within fifty light-years or so), we can use geometry. We measure the *apparent position* on the sky of a star relative to that of the much more distant *background stars* on a particular date, for example, February 1. Then, on April 1, two months later, when Earth has moved about one AU along the circumference of its orbit, we will again observe the subject star, finding that its apparent position relative to the more distant stars will have shifted slightly. We have all observed this effect—a tree that is close to us will seem to change its position relative to the trees in the distance if we slightly change our vantage point. This effect is called *parallax*. In the case of our distant star, we are talking here about extremely tiny shifts in its apparent position. Thus the parallax can be measured only by comparison to the positions of the distant objects within the field of view of the eyepiece of the observer's telescope. Knowing the parallax between two measurements and the length of the baseline separating the measurements, we can compute the distance to the object. The key to observing the parallax effect, therefore, is that the more distant stars don't noticeably change their relative apparent positions in the sky during the year, so they provide a fixed "coordinate system" for the measurement of the tiny shift in the position of the subject star.

The main problem in the measurement of the Earth-Sun distance is that there *is no* coordinate system painted on the sky that can be used for a measurement of the (angular) change in position of the Sun as we traverse a known baseline distance on Earth. The coordinate system provided by the distant "fixed stars" is visible only in a dark, nighttime sky. Bluntly put, the distance to the Sun can't be measured by parallax because the stars don't shine in the daytime! The trick to measuring the AU is not to measure the distance to the Sun but rather to measure the distance to Mars, for which the fixed stars provide a coordinate system for parallax, then to combine this with Kepler's laws of planetary motion to deduce the AU.

The AU was first measured to a precision of about 1 percent in 1685 by an experiment conducted by Giovanni Cassini, who was the chief astronomer in Paris. This required a long baseline, as afforded from the known diameter of Earth. Therefore one needed two measurements, at the same time, of the position of Mars relative to the fixed stars, separated by the diameter of Earth.

A naval vessel was given instructions to measure the apparent position of Mars while the ship sailed in the South Pacific. At the same time, the measurement of the position of Mars was made at the Paris Observatory. Once the ship returned, the two position measurements were compared, and the precise distance of Earth to Mars could be inferred using the known baseline distance between the two observers. Then, from (1) the known time length of Earth's orbital period (one year); (2) the known duration of the Martian orbital period (1.88 years); (3) the measured Earth-Mars distance at closest approach; (4) Kepler's laws of motion, which relate the orbital period to the radius of the orbit; and (5) a little algebra, one finally gets the distance from Earth to the Sun.[5] Finally, using Rømer's observed lag time of about 8 minutes for light to travel the distance from the Sun to Earth, the speed of light was determined to be 300,000 kilometers per second (186,000 miles per second).

It is useful to reflect upon how fast this is. This goes well beyond our normal everyday experiences of sights and sounds, leading us into a new world of physics. The diameter of Earth itself is about 12,720 km (7,904 miles) so the time it takes light to pass through this distance is about 1/24 of one second, approximately the limit of humanly perceptible time intervals. To traverse the circumference of Earth, light takes about 1/8 of a second. Such a timescale begins to become noticeable as the small time lag that we perceive while watching news reporters' dialogues, sent by satellite, from opposite sides of the globe. When the Apollo astronauts went to the Moon, however, we could readily hear the time delay in the exchange of conversation between the astronauts and Houston. Being about 384,000 kilometers away (240,000 miles; the lunar orbit is elliptical, and the distance varies by about 10 percent throughout a month), light signals take more than a whopping 2½ seconds to make a trip to the Moon and back. Rømer had discovered that the light we see from the Sun departed from its surface approximately eight minutes ago, whereas the light we see from the closest star, Proxima Centauri, took about 3.8 years to arrive; hence we say that Proxima Centauri is 3.8 light-years from Earth. The light from the typical brighter stars in the nighttime sky takes about 10 to 100 years to arrive at Earth, whereas for the most distant objects seen in the universe, light takes about 12 billion years to arrive. This is the distance to the *horizon* of our universe, since we are also seeing back to the time of the earliest stars and the very formation of the galaxies, and the beginning of our universe.

THE SPEED OF LIGHT AS SEEN BY MOVING OBSERVERS

The early measurements of the speed of light set in motion a debate that ultimately culminated two hundred years later in Einstein's special theory of relativity. The question was, What is being measured? Was Rømer measuring the speed of light emitted by Jupiter's moon Io? Or was this the speed of light emitted from the Sun, then bouncing off of the moving Io? Was the speed of light affected by the motion of Earth relative to Io?

Most scientists converged upon the idea that Rømer was measuring the speed of light as it propagated through something absolute, an unseen medium that fills the entire universe, an "ether," through which light travels just as sound travels through the air. The idea of a supporting medium, the ether, for the propagation of light derived from the Greeks but was a key idea resurrected in the era of Galileo. Yet having the capability of measuring the now finite speed of light opened a Pandora's box of new scientific questions. If light is indeed traveling through a static ether that fills all of space, and if we travel through the ether as we ride on planet Earth, could we detect Earth's motion by observing slight changes in the speed of light in different directions of space, or at different times of the year?

To obtain better control of any physical measurement, then and now, one ultimately wants, literally, to bring the measurement down to Earth— that is, to make the measurement in a laboratory located *on Earth*. In an Earth-bound laboratory measurement of the speed of light, one can locate the source and detector in a known, fixed reference frame of motion. The uncertainties and unaccountable effects of motion in planetary orbits— the effects of the source's or receiver's speed relative to the moving light and the difficulties of making precision measurements consistently throughout a calendar year—can then be eliminated. One has to be clever, since on Earth, the distance scales are smaller, and the measurement of timing becomes the problem, the problem of determining very short time intervals with high precision.

In 1850, two very skilled but highly competitive French scientists, Armand Fizeau and Jean Foucault, succeeded in making the first precise nonastronomical measurements of the speed of light on planet Earth. Fizeau was particularly interested in the question of the possibility of different values of the speed of light, dependent upon the observer's own state of motion or that of the source or reflector of light. If light was like a sound wave, traveling at a fixed speed in a material medium—the ether—then Fizeau hoped to see a different value of the speed of light as Earth moved relative to this medium. These scientists were, therefore, essentially in search of the ether.

Fizeau developed a mechanical timing apparatus, called a *stroboscope*, to measure the short transit time of light over a known distance in the lab. Foucault's method, which many students of physics have repeated in their studies, employed a light beam that reflected off a rotating mirror. The light beam subsequently reflected off a fixed second mirror, located more distantly, which bounced it back to the rotating mirror, which then reflected it onto a screen. In the finite time interval of the light's travel from the rotating mirror, out to the fixed mirror, and back, the rotating mirror had turned slightly. As the mirror rotated faster, the beam spot on the screen would be observed to shift its position. Measuring the shift in position of the beam spot on the screen, and knowing the distance from the rotating mirror to the fixed mirror, together with the rotation rate of the mirror, allowed a determination of the speed of light (see fig. 13). These techniques gave the speed of light with a precision of about ±0.5 percent.

The methods of Fizeau and Foucault were insufficient, however, to detect any finite differences in the speed of light due to the motion of Earth through an enveloping ether.

Albert A. Michelson was a diligent, young scientist working in 1877 at the US Naval Academy in Annapolis, Maryland. He devised refined versions of these stroboscopic techniques to achieve much more precise measurements of the speed of light. His first experiments, conducted when he was in his early twenties, were spectacularly successful, obtaining a value of 299,909 kilometers per second (186,355 miles per second), with a precision of ±0.02 percent, twenty-five times more precise than that of Foucault but still insufficient to see the effect of Earth's motion through the ether. With media coverage in all the major newspa-

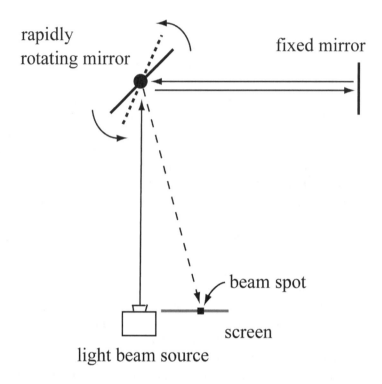

rapidly
rotating mirror

fixed mirror

beam spot

screen

light beam source

Figure 13. Foucault's rotating mirror experiment. The displacement to the right of the returned light beam spot on the screen, for a known rotation frequency of the mirror and distance to the fixed mirror, determines c, the speed of light.

pers of the day, Michelson became a celebrity for this high-precision experimental achievement, and he was inspired to devote his life to ultra-precise measurements of the speed of light, incorporating increasingly sophisticated techniques. Years later, teaming up with E. W. Morley, he developed an optical system that could finally, in principle, detect the effect of Earth's motion on the speed of light as it traversed the ether. After an earlier attempt in 1881 in Berlin, the refined experiment was performed in 1887 in the United States.

The procedure depended on what we now call a *Michelson interferometer*. This ingenious device simultaneously compares the transit time for light moving in two perpendicular directions. A beam of light is split into two beams, which then travel perpendicular to each other, reflect from mirrors, and are then recombined in an eyepiece. The wave nature of light implies that if the transit times differ by half a wavelength, the light waves will be seen to cancel each other out; and if they differ by one full wavelength, they will be seen to reinforce each other. The transit time difference depends upon the difference in the speed of light that the two beams have as they move along the different paths. Hence, one observes through the eyepiece the *interference pattern* of the recombined light beams. One then attempts to see variations in the pattern as the apparatus is rotated in position relative to Earth's hypothetical movement through the ether. In those days the entire apparatus had to float upon a bath of (highly toxic) liquid mercury to eliminate the effect of vibrations within the local environment. Such an experiment could never be conducted today in a typical university setting in the United States with modern EPA rules.

And what did the Michelson-Morley experiment detect? Nothing. The experiment yielded a null result! *No difference* in the speed of light, traveling with the motion of Earth or perpendicular to it was found. This null result was a spectacular kick in the shins for the ether theorists. The experiment created a complete conundrum. How is light propagating? Why does it violate the commonsense expectations that both Galileo and Newton had? What is going on?

To appreciate the shock value of the Michelson-Morley experiment, consider two young physicists of the future, named Jackie and Hillary, who have all the latest equipment. Each of them happens to be carrying the suave new Acme pocket silicon strip matrix speed-of-light detector, with subnanosecond precision and an iridium laser with a built-in helium atomic clock. Jackie stands on the platform of the Maglev Train Station, while Hillary rides by on an express train traveling at one-half the speed of light.

At the instant that Hillary's window on the train passes Jackie on the platform, the ladies have arranged for a flash of light from a flash bulb on the platform. Jackie measures the speed of the photons of light coming from the flash with her detector, and Hillary measures the speed of light from the same flash on the train. Later, the two get together for coffee. "Say, Jackie, what did you measure for the speed of light the other day when I passed you in the express train as you were standing at the train platform?" asks Hillary.

"Why, exactly $c = 299{,}792{,}458$ meters per second, the usual speed of light," replies Jackie. "What did you measure?" "Hmm, that's strange," says Hillary, "my Acme detector was working perfectly, yet I measured

c = 299,792,458 meters per second, the same as your result, the usual speed of light, within the precision of ±1 meter per second." She continued, "Yet I was traveling in the train at half of the speed of light relative to you. I am astonished that I have measured a value of the speed of light that is identical to yours! How can this be?"

Indeed, both observers measure the same exact speed from the same flash of light. There is no Galilean addition of the speed of the train, which is carrying one of the observers. These detectors were so precise (much more so than Michelson and Morley's) that they should have shown a significant discrepancy at the relative speed of a moving high-speed train. What indeed is happening?

The Principle of Relativity

As we have noted, Galileo discovered the principle of relativity: *all states of uniform motion, called inertial reference frames, are equivalent for the description of physical phenomena*. As we change our state of motion and end up in any different state of motion, the laws of physics appear to be the same to us. The principle of relativity is a continuous symmetry of the laws of physics—we can continuously change our state of motion from any one to any other.

Consider an astronaut in ultradeep space. In fact, let us suppose that the hapless astronaut is way off course and has now gone infinitely far away from all points of reference, such as other stars or galaxies or visible objects. Let us assume that there is no gravity, no cosmic rays, and no radiation in the universe left over from the big bang—nothing that the astronaut might measure to determine his state of motion. The astronaut is marooned in completely empty and dark space.

As the astronaut drifts through space, everything else in the space capsule—the food tubes, the space helmet, the souvenirs, and all the other objects—are at rest *relative to the astronaut*. The principle of relativity states that there is *no experiment that the weightless astronaut can do* to detect his state of motion. Should the astronaut apply his thrusters and accelerate, he will feel himself pushed back in his seat. After turning off the thrusters, he will again be inertial and weightless. There is *no experiment that the astronaut can do* to detect a difference in the laws of physics between *his original and his new state of motion*. All that anyone can ever do is detect *relative motion*, motion relative to some marker or reference system such as Earth, the Sun, or a distant star such as Alpha Orionis, or anything—hence the name, *relativity*—but there are no such markers in completely empty and dark space.

This situation is similar to that of rotational symmetry. There is no absolute up or down, or sideways, or back and forth in the universe. We can always tell how something is rotated *relative to something else*, but there is no absolute orientation of anything in the universe. Just as we can perform a transformation such as a rotation, which rotates an object from one orientation to another, we can also apply a transformation that changes the object from one state of motion to another, such as applying the rocket thrusters of an astronaut's vehicle.

We call the transformations that change the velocity of a system from one value to another *boosts*. We can boost an object in any direction in space—in essence, we can give it a boost—to any velocity we want, according to Galileo. When the astronaut fires his thruster rockets, he is performing a boost transformation on himself and his space capsule. The invariance of a physical system, or the laws of physics, to a boost is there-fore a symmetry operation, much like rotating a sphere is a symmetry operation.

But Galileo also had another fundamental concept in mind—the "principle of absolute time": *all observers, no matter how they move through space, must conclude that the time interval between any two events is the same.* This principle is a statement about the symmetry of time—time would be invariant under boosts, according to this principle. The concept of absolute time was fundamental to all of physics from Galileo to Einstein. However, it is, as we'll see, the key piece of baggage that Einstein discarded—the principle of absolute time is *wrong!*

The Overthrow of the Relativity of Galileo

The physical world is a fabric of events. Events are things that happen at precise positions and times in space and time. Given any pair of events, knowing their coordinates, we can compute the separation between them, L, and the time interval between them, T. For example, if both events occur on an imaginary x-axis, one event occurring at x_1 and the other at x_2, then the separation is $L = x_2 - x_1$. Likewise, if our clock says that event 1 occurred at t_1 and event 2 at t_2, then the time interval is defined by $T = t_2 - t_1$. Now suppose another observer sees the same events but is moving relative to us with a velocity v. We'll assume that the observer is moving in the direction from the first event, 1, toward the second event, 2. The question is, what will the moving observer measure for the distance and time interval between the events?[6] Galileo said that the answer is a Galilean boost:[7]

$$L = L' - vT, \quad T' = T.$$

This is called the *Galilean transformation*. The second equation is just the mathematical statement of the *absoluteness of time*. The first equation shows how the comparison of the distance separation measurement between two events is affected by the relative motion. The Galilean transformation is a continuous symmetry because the speed, v, by which we boost can vary continuously. It is not hard to show that the Galilean transformation implies that the speed of anything, *including light*, changes when we chase after it. A Galilean boost, moreover, can be performed for any velocity v. There is no upper limit to the relative velocities of two observers in classical physics, and it can be many times greater than the speed of light.

If my cat, Ollie, is in an inertial reference frame fleeing me at a very high rate of speed with my pet hamster, Arlo, in his mouth, then I will attempt to boost myself into a reference frame in which I can overtake Ollie to recover Arlo. If Ollie is traveling at a velocity v away from me, and if I boost myself to a velocity v' in the direction of Ollie, then I will observe Ollie traveling at velocity $v - v'$. By choosing a v' that is large enough, I can overtake Ollie and hopefully rescue poor Arlo in time. This is all theoretically permitted in the physics of Galileo and Newton. And this is in accord with my everyday experience.

The revolutionary significance of the Michelson and Morley experiment was monumental. It showed, however, that $c' = c$, no matter how fast we travel after a light signal. This result was indeed so shocking that it might well have brought Galileo back from the grave. It is in fact impossible to reconcile it with Galileo's form of the transformations between inertial reference frames. It is also seemingly paradoxical.

Consider again the case of Ollie fleeing at a very high rate of speed, with Arlo in his mouth. If Ollie somehow managed to have a speed equal to that of light as he runs away from me, then no matter how fast I run after Ollie, I can never catch him, or even change the speed at which he is receding from me! Hence, I can never hope to rescue poor Arlo. We evidently have a paradox, namely, $c - v = c$ for any v! But how can this be true? The laws of nature must be mathematically consistent, and this result, which is like saying $4 - 3 = 4$, is seemingly absurd.

Some physicists attempted to argue that there really was an ether, but subtle dynamic effects associated with travel through the ether modified the results in a way consistent with the experiment of Michelson and Morley. Hendrik Lorentz and George Fitzgerald discussed the idea that all physical objects were dragged by the ether such that their lengths were shortened or contracted in the direction of motion. This would also cause clocks to run slow, and it led to a "conspiracy" in which the moving observer would always measure the same value of c, no matter what her speed. This was a misguided rationale, and its underlying logic was really an attempt to rescue the ether, but it was the beginning of modern special relativity.

THE RELATIVITY OF EINSTEIN

It was Albert Einstein in the early years of the twentieth century who resolved the conundrum. In 1905, with some deceptively simple broad strokes, the twenty-six-year-old patent office clerk in Bern, Switzerland, accustomed to thinking while pushing a baby stroller, brought down the whole house of Galilean and Newtonian classical physics. His new concept of space and time completely overhauled our understanding of nature and led to modern physics. It represented, and still represents, one of the most stunning achievements of the human mind. It was based entirely upon thinking about nature from the perspective of symmetry.

Einstein was led to special relativity by thinking in terms of the symmetry principles that defined light, as it was understood in the late nineteenth century. In fact, this is, in a sense, the greater import of Einstein's vision. It was Einstein who radically changed the *way* people thought about nature, moving away from the mechanical viewpoint of the nineteenth century toward the elegant contemplation of the underlying symmetry principles of the laws of physics in the twentieth century.

Einstein made the basic assumption that we will always observe light to travel at the same fixed speed, no matter how fast we attempt to chase after it.[8] Put into the language of symmetry, *the speed of light is invariant for all observers*. Recall that a symmetry is something that is invariant under a transformation. Einstein is demanding that the speed of light be invariant under the transformation of boosts (whereas previously, Galileo had demanded that time intervals were the same for all observers). Einstein's special theory of relativity is therefore defined by two principles:

- The principle of relativity: *All states of uniform motion, called inertial reference frames, are equivalent for the description of physical phenomena.*
- The principle of the constancy of the speed of light: *All observers will obtain the same value for the speed of light in any inertial reference frame.*

The first principle is simply borrowed from Galileo. However, the second principle is the result of the Michelson-Morley experiment, now imposed as a *new* symmetry principle upon nature. We now discard Galileo's implicit notion of the absoluteness of time. Einstein demanded that these two results must be true and must coexist faithfully. Incidentally, it is possible that Einstein, who focused upon the symmetry inherent in the mathematical theory of electrodynamics, was not influenced by the Michelson-Morley experiment and may not have been aware of it at the time he founded special relativity.

These principles of special relativity can be viewed in an elegant and succinct way. The symmetry of special relativity involves a totally new geometric concept of the "distance" between any two events. This new distance is called the *invariant interval*. The invariant interval involves the separation in *time* between any two events as well as their separation in *space*.

Consider any two events, 1 and 2. The two events have length separation L and time separation T in a given reference frame that we'll arbitrarily call the "rest frame." The invariant interval between the two events is denoted by the Greek letter τ (tau), and it is defined by a simple formula: $\tau^2 = T^2 - (L/c)^2$. This formula bears a striking resemblance to the Pythagorean theorem of geometry. Given a right triangle with *sides* of lengths x and y, the length of the *hypotenuse* of the triangle is z, where $z^2 = x^2 + y^2$ (or "the square of the hypotenuse equals the sums of the squares of the other two sides"; this is something film buffs will recall was quoted when the Wizard of Oz conferred a testimonial, not a brain, upon the Scarecrow). Einstein's special theory of relativity actually proposes a new kind of geometry of space and time, now called space-time, in which the hypotenuse is the invariant interval, τ, and the legs of the triangle are the time separation between events, T, and the space separation L, divided by the speed of light, c. But there is one new, and very significant twist to Einstein's geometry: the space part, $(L/c)^2$, enters with a minus sign into the new Pythagorean formula, while the time part, T^2, enters with the normal plus sign. This happens, because, as we know from experience, time is different from space.

Now, different observers, moving relative to the rest frame with a velocity v, will measure a different time interval, T', between the two events, as well as a different space interval, L'. However, Einstein's new symmetry is that the invariant interval between any pair of events, τ^2, is *the same* for all observers, no matter how they are moving. That is, if we compute τ^2 using L and T, we'll get exactly the same result as using L' and T'. Indeed, we can essentially combine Einstein's two new defining principles of relativity into one single, powerful symmetry principle: *the invariant interval between any two events must be the same for all observers, no matter how they are moving relative to one another.*

If the two events happen at the same point in space, then the spatial separation in the rest frame between the events is $L = 0$. Thus the invariant interval is simply $\tau = T$. So the invariant interval is the *actual time elapsed on a clock* relative to which the events occur at rest. We often refer to the invariant interval by another name: the *proper time interval* between the events.

On the other hand, if the two events in space-time are connected by a light signal, such as a flash of light from a source event to a receiver event, then the invariant interval, or proper time interval, between them is zero, $\tau = 0$. Since this is the same for all observers, all observers will therefore conclude that the speed of light is the same, no matter with what velocity they are moving.

Einstein therefore asked, "What form of boosts would leave the invariant interval, τ, the same (invariant) for all observers?" What he found, when applied to the situation of observers moving at velocity v in the direction of event 2 and away from event 1, was that their observed time interval, T', and spatial separation, L', are related to L and T in the rest frame by what we will call "Einstein boosts":[9]

$$L' = \gamma \cdot (L - vT), \qquad T' = \gamma \cdot (T - vL/c^2).$$

Here the new mathematical factor, called gamma or sometimes the *Lorentz factor*,

$$\gamma = \frac{1}{\sqrt{1 - v^2/c^2}},$$

permeates everything in special relativity.

Formulas can be painful things to look at—and even more painful to manipulate. Nevertheless, it is not hard to deduce, using these formulas and a little high school algebra, that $T'^2 - (L'/c) = T^2 - (L/c)^2$.[10] This confirms that the invariant interval, or the proper time, is therefore the same for *both sets of observers* under an Einstein boost. Einstein engineered his boost formulae to achieve this very goal. Einstein's boosts are the correct symmetry transformation of time and space intervals between different observers that are moving relative to each other, and they therefore replace Galileo's.

Einstein boosts differ from those of Galileo in two very significant ways. First, there is the ubiquitous γ, the "gamma" or "Lorentz factor," which is necessary to make the interval between two events invariant. It guarantees that a light flash is always seen to spread out spherically into space at the speed of light, for any observers, no matter what his velocity v. Second, we see that time is no longer absolute. Time and space are commingled when we move relative to one another, and time ceases to be absolute.

Furthermore, we can see that for low speeds (that is, when v is much less than c), the Einstein boosts approximate the form of the Galilean boosts, $L = L' - vT$ and $T' = T$. For small velocities, the discrepancies between pure Galilean boosts and Einstein's boosts are extremely small; therefore special relativity becomes an unobservable correction for slowly moving objects. Viewed yet another way, if we take the speed of light to be infinite, we see that Einstein's boosts also predict $T' = T$, and we again recover the absoluteness of time! For slowly moving observers, this shows that Galilean relativity was a completely reasonable approximation. The absoluteness of time and the Michelson-Morley experiment could only be totally reconciled, though, if the speed of light were infinite!

We can think of the invariant interval between two events as the length of an Acme space-time classroom event-pointer. The event-pointer begins with its handle at one space-time event (1) and ends with its "tip" at another space-time event (2). Einstein boosts resemble a kind of "rotation" in space-time, and the length, that is, the *invariant interval* of this event-pointer, remains the same, just as ordinary rotations in space preserve the length of an ordinary classroom pointer. In this sense, boosts, which produce different states of motion, are similar to rotations in ordinary space.

Historically, the mathematical form of the boosts is something that Lorentz had essentially derived a few years earlier with the idea that the ether produces a drag on physical objects. For this historical reason, Einstein boosts are called the *Lorentz transformations*. But the ether simply does not exist. Today we view Lorentz transformations (Einstein boosts) as true and correct symmetry transformations of the laws of physics under motion—the symmetry that respects the two defining principles of Einstein.

THE BIZARRE EFFECTS OF SPECIAL RELATIVITY

What now emerges from special relativity is truly bizarre.

Suppose two objects are at a distance L apart in their rest frame. How far apart are they, as seen by moving observers? With a careful analysis of the measurement of the length of an object between two objects (one must measure the positions of the endpoints simultaneously), we would find that the observed distance between them is, $L' = L\sqrt{1 - v^2/c^2}$.[11] The moving observers measure the distance between the objects to be *contracted*, or shortened, by a factor of $\sqrt{1 - v^2/c^2}$. Hence, if the speed is approaching that of light—for example, $v = 0.866c$—the distance appears to be half as long as it is in its rest frame.

In relativity, moving objects are observed to be contracted in their direction of motion, becoming squashed flat like pancakes as they approach the speed of light. For example, suppose the object is a proton, normally a spherical blob of quarks at rest. After it is accelerated at Fermilab, so that it is moving at 99.99995 percent of c, then we observe that it is squashed like a pancake in the direction of motion by a factor of 1/1000. In fact, the faster the object travels, the shorter and shorter it becomes in the direction of motion, until at $v \rightarrow c$ it has no length (in the direction of motion) at all! Objects become squashed into pancakes in their direction of motion as seen by stationary observers, but if we ride along with the object we notice no effect. In fact, paradoxically, we would then observe the universe, as we look out our relativistic spacecraft window, to be moving near the speed of light in the opposite direction. It is then the universe that appears squashed like a pancake relative to us!

Suppose we have a clock that simply flashes a repeating light signal every $T = 1$ second. The flashes of light serve as a "tick" and "tock," like a metronome used by piano students. What would moving observers measure for the tick-tock interval?

It's not hard to see that they would observe the flashes at regular intervals, but with time interval $T' = T/\sqrt{1 - v^2/c^2}$ (set $L = 0$ in the Lorentz transformation for the time interval). The moving observers would thus conclude that the time interval between flashes, T', is longer than $T = 1$ second. That is, the clock would apparently be running *slow*! Of course, from the observers' point of view, it is the clock that is moving, relative to them, with a speed v in the opposite direction. Therefore all moving clocks relative to any "stationary observer" are seen to run slow.

For example, if the moving observers have the enormous speed of $v = 0.866c$, then they would observe that $T' = 2$ seconds. This means that they observe that the clock is running at *half* its normal speed. The clocks of any system moving at speeds approaching the speed of light, relative to us, are observed to run slow. This phenomenon is called *time dilation* (a term mentioned earlier). Observing a system approaching the speed of light, we see the tick-tock interval become infinite, and the clock apparently ceases to run at all.

Indeed, we observe in the lab that elementary particles traveling near the speed of light actually live longer than they do at rest—their half-lives are lengthened in complete accord with the predictions of relativity. But if we could ride along with the relativistic particle, we would observe no time dilation, and it would be the universe, moving by in the opposite direction, all of whose clocks would be slowed down as seen by us!

One of us (LML) received his PhD for verifying the prediction that clocks observed to be moving near the speed of light appear to run slower. The "clock" was a beam of muons, particles that decay in about 2.2 microseconds (i.e., 2.2 millionths of a second) when they are at rest. The muons were produced as by-products of collisions involving protons accelerated by the Columbia University synchrocyclotron. A beam of muons traveling at 86 percent of the speed of light was measured to have a half-life of 4.2 microseconds, about twice the half-life of when the muons were at rest. This change in muon half-life was measured to about 5 percent precision way back in 1950. Muons produced today at Fermilab, traveling much closer to the speed of light, have half-lives that can be a thousand times longer than the resting lifetime. Please don't envy a muon its lengthened life span—from its point of view, your clocks are running slow, and you have the lengthened lifetime!

Time dilation leads to a famous brainteaser called the *twin paradox*.

We imagine that a bride, after her long and romantic honeymoon, puts aside her personal life for the good of science and signs up for a daring space mission. After a long hug, the wife says farewell to her husband and blasts off from Earth; her parting words are, "Honey, I'll only be gone for two weeks." She travels to a distant star at nearly the speed of light. Though the star is ten light-years away, from the vantage point of her reference frame the distance is length-contracted to a mere one light-week. Upon arriving at the star, she takes some pictures, immediately reverses her thrusters, and heads home at the same high speed. Indeed, she has been gone for only two weeks, according to her clock. When she gets home, she rushes to the arms of her husband.

For the husband, who stayed home and faithfully tracked her long voyage, the round-trip has taken twenty years, and he is the worse for the wear of time, having aged twenty years. Furthermore, he had noticed that, during the trip, she was traveling so fast that her clocks were virtually *frozen* by time dilation such that the elapsed time for the trip, according to the clocks carried along by her ship, was only two weeks. When they meet upon her return, indeed he has aged twenty years, and she two weeks. Like wine, however, this did not dampen the enthusiasm, even though it slightly diminished the performance.

The paradox of this situation arises when we consider this in greater detail from the wife's perspective. Indeed, she was at rest in her reference frame, observing her husband to be moving in the opposite direction, and it is therefore his clocks that should have slowed down, as seen by her. So how did the wife interpret the result that her husband has aged so much, and she has not? The trick is that, from her perspective, this could not be resolved without considering the effect of *acceleration*. She has experienced (extreme) acceleration to get her near the speed of light, while the faithful husband has had no such experience. During this time she has seen the enormous distance to the star shrink by length contraction from ten light-years down to the mere one light-week, as the star is then approaching her at nearly the speed of light. It is during this "boost phase" that her husband, from her perspective, did all the aging. It is as though he had been inertially, or freely, falling in space, while she had been accelerating, as if caught in a strong gravitational field.

This anticipates a fact that Einstein later recognized: inertial clocks in strong gravitational fields must run slower than clocks in free fall. Indeed, this anticipates a result in general relativity called the *Einstein redshift*. Light emitted from the surface of an enormous star, where the gravitational force is very strong, is actually observed to be *redshifted*, as though the atoms on the surface of the star have clocks running slower than observers in free fall far away (red light has a wave with a lower frequency than blue light). Thus, during the acceleration phase(s) of the wife's trip she observed that her husband aged about twenty years, whereas she, the accelerating one (effectively in a strong gravitational field, being Einstein redshifted) didn't age at all. The twin paradox is resolved—it is not a paradox after all. And she is delighted by the fact that she has not aged more than two weeks during the entire trip.

The essential reason for these weird effects is that two events that are simultaneous for one observer are not generally simultaneous for another. This is the hallmark of special relativity and underlies all of these bizarre effects.

ENERGY AND MOMENTUM IN SPECIAL RELATIVITY

Einstein's reaction to the puzzle of the speed of light was to accept the validity of the constancy of the speed of light for all observers and to throw out the absoluteness of time. Moreover, *all* the laws of physics must have this symmetry—they must be invariant under boosts. The Lorentz transformation completely replaced Galileo's transformation between moving observers. Therefore, all of the old physics that was based upon Galilean relativity, such as Newton's laws of motion and theory of gravity, needed to be modified.

Since none of Newton's equations were invariant under Lorentz transformations (as the velocity approaches the speed of light), Einstein confidently assumed that they must all be wrong, despite over 250 years of successful applications! He therefore began to think about what had to be done to repair the old Newtonian concepts, such as force, momentum, angular momentum, and energy, with the new correct and relativistic ones. Einstein was guided by two ideas: whatever was changed, it was necessary that the validity of Newton's laws was recovered for physics at low velocities. Moreover, the symmetry of relativity must hold for all the new laws of physics. We do not know at what point Einstein realized, in his chain of thoughts, that this would radically and forcefully change the future of the human species.

In special relativity, we have seen that the defining symmetry principle is that the invariant interval between any two events will be observed to be the same for all observers: $\tau^2 = T^2 - L^2/c^2$. Space and time now enter quite symmetrically into this formula, like the Pythagorean formula for the hypotenuse of a right triangle, both raised to the second power.

What about energy, momentum, and mass? These quantities are related in Newton's classical physics, but we must find a new relationship between energy and momentum that now holds in special relativity. The symmetry between space and time suggests a corresponding symmetry between energy and momentum. In fact, we can appeal to the correspondence implied by Noether's theorem.

Let's consider a particle with energy E, momentum p, and mass m. We recall from Noether's theorem that time is related to energy, $T \leftrightarrow E$, and also that space is related to momentum, $L \leftrightarrow p$. This suggests that, in special relativity, there must be a corresponding "Pythagorean formula" involving energy and momentum. Indeed, we might suppose that the corresponding quantity, $E^2 - p^2 c^2$, much like the invariant interval, $T^2 - L^2/c^2$, can *also* be considered to be invariant under a Lorentz transformation. This means that an observer in the "rest frame" measures a particle to have energy E and momentum p, while a relatively moving observer will measure a different energy, E', and momentum, p'; nonetheless, the symmetry of relativity will require that $E^2 - p^2 c^2 = E'^2 - p'^2 c^2$.[12] (Here we have put the factors of the speed of light, c, in the right places to get all the units to work out correctly. Recall that energy has the units of momentum times velocity, pc.)

Of course, the inertial mass of the particle, m, must somehow enter the new formula that relates the energy and momentum of a particle. This is because the inertial mass of an object is something intrinsic to it and should be an invariant quantity as well. Einstein thus guessed that this new invariant formula involving momentum and energy must be *equivalent to the inertial mass, m*: $E^2 - p^2 c^2 = m^2 c^4$. (Again, we have to include the factor of c^4 on the right-hand side to get the units right. This would not be a problem if we simply used sensible units in which $c = 1$!). Let us understand the significance of this remarkable result. What happens if our particle is at rest? In that case, the momentum is zero, $p = 0$. Our formula thus becomes $E^2 = m^2 c^4$. But to get the energy, we must take the square root of both sides of this formula. And, when we do that, we find the following result: $E = mc^2$.

Now, before shouting "Eureka!" we have to check one other thing. What if the particle is moving and has a very small momentum? From his new formulas Einstein found that, if the momentum is small—meaning small compared to mc—then the energy is[13]

$$E \approx mc^2 + \frac{p^2}{2m} + \ldots .$$

The second, additional term on the right is exactly the kinetic energy of a slowly moving (compared to the speed of light) particle of Newton (or equivalently,

$$\text{K.E.} = \frac{1}{2} \, mv^2,$$

as we have seen previously, when we computed the energy of a moving automobile, since the momentum is just $p = mv$).

This must have elicited the grandest of all possible eurekas heard since Archimedes first uttered this exclamation! This result tells us something profound: *if a particle is at rest we still have energy*, as demonstrated by the resoundingly famous equation

$$E = mc^2.$$

The implications of this formula are earth-shattering. Inertial mass is equivalent to a certain amount of energy. This equation is so famous that it regularly makes appearances on T-shirts, on license plates, in cartoons, in Hollywood productions, on subway and restroom walls, in Broadway musicals, on doodles on ink blotters in the Oval Office, and throughout countless other venues. Although mass and energy are different things, this simple formula informs us that one can in principle be converted into the other, and vice versa. This formula literally unleashes all of the energy in the universe, for better or worse.

Suppose we could convert one kilogram (about 2.2 pounds) of mass into energy? Einstein's formula says that we'll get from this $(1 \text{ kg}) \times c^2 = (1 \text{ kg}) \times (3 \times 10^8 \text{ m/s})^2 = 9 \times 10^{16}$ joules of energy. This is an enormous amount of energy, able to make a 10,000-kilogram (about 10-ton) spacecraft travel with a velocity of more than 1 percent of the speed of light. The Einstein energy-mass equivalence furthermore tells us that the mass of the nucleus of uranium-235 is actually greater than the masses of the daughter nuclei and the free neutrons that are produced from the disintegration. Without taking account of the conversion of rest energy into the final energy of radiation, we would never see the overall conservation of energy in the process. Any process in which a conversion of mass to energy occurs, that is, the total inertial mass is not conserved, is a process that can be described only in Einstein's special theory of relativity.[14] This is the formula that is most identified with the age of nuclear physics. However, it is a formula that holds for all things throughout the entire universe for all time.

GENERAL RELATIVITY

Special relativity also requires a new theory of gravity to replace that of Newton. Newton's theory of gravity cannot be correct, since no signal can ever travel faster than the speed of light, but Newton's theory predicts that the force of gravitation propagates instantaneously between two objects. Newton's theory can only describe slowly moving particles and systems, particles that are *nonrelativistic*, involving particles and processes in which the rest energy converting to energy of motion does not happen. The full theory of gravity is Einstein's general theory of relativity, and it is an intellectual masterpiece. As we have hinted previously, it involves the principle of inertia in a deeper and more fundamental way.[15]

Let's jump ahead and ask a simple question that anticipates one of the most major and dramatic results of Einstein's general theory of relativity. What happens if a particle attempts to escape the surface of an object with such a strong pull of gravity that it would require the particle to convert *all of its rest energy* into motion in order to escape?" Indeed, the massive object would then forbid the escape of the hapless particle, since there would be nothing left of the escaping particle once it managed its escape.

It turns out that escape from a massive object is impossible if all of its mass, M, is compressed down within a radius of $R = 2G_N M/c^2$, where G_N is Newton's gravitational constant.[16] The object has then become a *black hole*. R is called the *Schwarzchild radius* of the black hole. Any object of a given mass, M, whose radius is smaller than R, as determined from this formula, will be a black hole. No particle, not even light, can escape a black hole from a distance within the Schwarzchild radius of the black hole. For example, if Earth were the object, and we put in the appropriate numbers, we find that to have the properties of a black hole, Earth would have to be compacted down to the miniscule radius of $R = 2G_N M_{Earth}/c^2 = 8.9 \times 10^{-3}$ meters, or about a quarter of an inch. This means that compressed to this scale, Earth would become a black hole. For the Sun, the Schwarzchild radius is about two miles. The density of the matter filling a region this size would grossly exceed that of the atomic nucleus. Today it is widely believed that the centers of most galaxies contain humongous black holes, with masses that are many millions of times greater than the mass of the Sun.

General relativity explains gravitation as a curvature, or bending, or warping, of the geometry of space-time, produced by the presence of matter. Free fall in a space shuttle around Earth, where space is warped, produces weightlessness, and is equivalent from the observer's point of view to freely moving in empty space where there is no large massive body producing curvature. In free fall we move along a "geodesic" in the curved space-time, which is essentially a straight-line motion over small distances. But it becomes a curved trajectory when viewed at large distances. This is what produces the closed elliptical orbits of planets, with tiny corrections that have been correctly predicted and measured. Planets in orbits are actually in free fall in a curved space-time!

Newton's theory of gravity is ultimately only an approximation to Einstein's theory in the limit of small velocities of motion compared to the speed of light. General relativity correctly accounts for residual anomalies in the planetary motions, such as the fact that Mercury's perihelion (the location of the distance of closest approach to the Sun) advances about one degree per century, an effect for which Newton's theory cannot account. General relativity also correctly predicted the bending, "lensing," and color shifting of starlight as it passes or leaves gravitating objects. Einstein's general theory of relativity applies to the universe as a whole and correctly predicts that it should be expanding, that space is literally being created. And, as we have seen, general relativity predicts that objects can become so massive that they can trap all matter and light from ever escaping from their surfaces, becoming black holes, nature's answer to the epic question, what is Tartarus, or hell, like?

chapter 8

REFLECTIONS

Now if you'll only attend Kitty and not talk so much, I'll tell you all my ideas about Looking-glass House. First, there's the room you can see through the glass—that's just the same as our drawing room, only things go the other way. I can see all of it when I get upon a chair—all but the bit behind the fireplace. Oh! I do so wish I could see that bit!

—Alice, *Through the Looking Glass*

W hen Alice climbed up on the mantle of her Victorian parlor to get a better view and see if there was also a fire in the fireplace of the "Looking-glass House," she tumbled into a new world. In this world the normal laws of physics were suspended—chess pieces muttered and roamed about the countryside, Humpty-Dumpty took a great fall, and "all mimsy were the borogoves, and the mome raths outgabe."

Yet, you may ask at a hypothetical level, what kind of physical world do we really see in a mirror? Indeed, we see a different world, alphabetical letters reversed, the sunlight entering windows into a room that almost looks the same as ours, yet new, and our own image, as we are accustomed to it, but not as others see it, with that freckle and the part of the hair on the wrong side, but more or less the same. It all comes down, ultimately, to one thing: "things go the other way," as Alice said—left and right are reversed.

This left-right–reversed world through the looking glass is otherwise hardly changed at all. Were we astute observers of everything that happened in this world, as methodical as Kepler in trying to understand its rules and laws, what would we conclude? Would we find any difference between the laws of nature in the mirror world, compared to the laws of nature in our world? Or is the "dual" world through the looking glass equivalent to ours in its most fundamental laws of physics? Would we find this to be a symmetry—to enter into the "Looking-glass House"—to find in that world that only the seemingly most superficial things, left and right, are reversed, yet the laws of nature otherwise remain the same?

As we have seen, not all symmetries are continuous. Noether's theorem strictly applies only to the continuous symmetries, yet even discrete symmetries can lead to certain kinds of conservation laws (particularly in the realm of quantum theory). Discrete symmetries play a fundamental and mysterious role in nature as well as continuous ones. Our world is full of discrete symmetries. Could the reversal of left and right be such a symmetry of nature?

REFLECTION SYMMETRY

Look at a particular physical system, such as the Taj Mahal, shown in figure 14*a*. Here we see the familiar view of the front facade of the Taj Mahal, with its majestic reflection pool. We can use the picture to illustrate the concept of a discrete symmetry operation, or transformation, known as a *reflection transformation*.

In the second picture, figure 14*b*, we have drawn a line down the center of the Taj Mahal facade. This is the "symmetry axis" of the picture of the Taj Mahal. Using a computer graphics program we have "reflected" the first picture about this line. This means that we take every point, such as point *x* on the left of the line, and interchange it with point *y* on the right, where *x* and *y* are equidistant from the symmetry axis. The reflection of the picture is a two-dimensional transformation, but we can actually imagine performing the reflection operation on the full three-dimensional object itself. Then the symmetry axis becomes a plane containing the vertical line. Every point *x* to the left of the plane is then swapped for an equivalent point *y* to the right of the plane, where the line connecting each pair, *x* and *y*, is perpendicular to the symmetry plane.

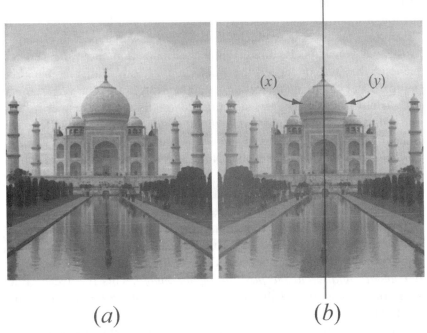

$$(a) \qquad\qquad\qquad (b)$$

Figure 14. The Taj Mahal, before (*a*) and after (*b*) reflection about the symmetry axis. (Photo by CTH.)

Physically the Taj Mahal appears the same after this transformation. A mathematician would say that the Taj Mahal facade is *symmetric under the reflection transformation* or that the facade *possesses a reflection invariance*. The reflection operation reverses left and right. Any point on the left of the symmetry axis of the Taj Mahal has been *mapped onto* an equivalent point on the right of the symmetry axis by the transformation, and vice versa.

The reflection operation is what we see in a mirror, and an image of a reflection transformation of an object can be obtained by photographing the object as seen through a mirror. For example, if we turn our back to the Taj Mahal, face a mirror that reflects the image of the facade of figure 14*a*, and photograph the mirror image, we will get the same image as the reflected Taj Mahal facade of figure 14*b*.

The reflection symmetry has been used by the architects to invoke a sense of perfection, a sense of divinity and beauty, in the design of the Taj Mahal. Art imitates nature through the incorporation of discrete symmetries, and indeed, reflection symmetry can be found throughout nature. In anatomy, the human body and the human brain itself, to a good approximation, are *bilaterally symmetric*. Hence, when you look at yourself in the mirror you look pretty much the same way as you appear to someone looking at you. That is, doing a reflection operation about the vertical plane of the face, or indeed the whole human body, will give back an approximately equivalent face, or human body. A brain removed from its skull is physically symmetric about the central fissure that defines its left and right sides. The left brain and right brain typically function differently in an organism, but in terms of shape and structure (anatomists say *morphologically*) they are the same. Numerous organisms have reflection symmetries of different kinds.

Although many things are invariant under reflections, that is, "they are mapped back into the same thing," there are also many things that are not invariant under mirror reflection. For example, our left hand, under reflection, becomes a right hand. Right and left hands are distinct from one another. This happens because there is a *sense in which we can curl our fingers relative to the position of our thumb*. This relative curling sense of the fingers and the placement of the thumb defines left- and right-handedness.

Imagine a species with a hand having two thumbs, one where the little finger should be, together with the normal thumb, such that the hand is symmetrical about the middle finger, as in figure 15. In this species there is no difference between left and right hands. Imagine the problems with traffic on the alien's planet—who would have the right-of-way? How would we distinguish it from the "left-of-way"? Hitchhiking would also have its challenges.

However, for humans there is a difference between left and right hands. If we take a box of gloves that have padded palms (something to define the difference between the top of the hand and the palm), we can always determine which glove goes with the left and which goes with the right hand. For the alien species there is no difference between a right glove and a left glove. We thus see a mathematical property of reflections and its consequence for the physical world. Left and right are either the same thing, or they are different and form a pair of mirror image partners. For an alien, the mirror image of his hand is the same as his hand itself. We would say that under reflections the alien hand is a *singlet*. For humans we say the hands form a *doublet*. In the case of our human hands, there can be only, at most, two partners under reflection. There is no third partner to a reflection operation, because if we do a reflection operation

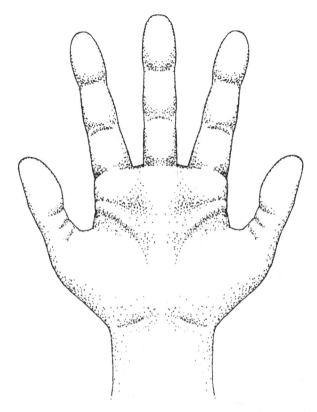

Figure 15. Hand of an alien species with two thumbs. The hand is neither a right nor a left hand. (Illustration by Shea Ferrell.)

after a previous reflection operation on our right hand, we'll get back our right hand (mathematically, the "square of a reflection"—that is, a reflection times a reflection—is the identity). If we reflect left we get right, and vice versa. We refer to something that is not invariant under reflections, something that changes into something else under reflections (like the right hand changing into the left hand) as having *handedness*.

It's not hard to make physical things with handedness. A box of screws from a hardware store will usually be "right-handed." This means that a rotation of the screwdriver in the same sense as the curling of the fingers of the right hand moves the screw forward, in the direction of the thumb. Seen in the mirror, the right-hand rotation becomes left-handed, but the mirror image of the screw still moves forward, so, the mirror-image screw is "left-handed." The point is that a left-handed screw can just as easily be manufactured, and it is completely compatible with the laws of physics—nothing violates the laws of physics to make a left-handed screw, it just takes a special order from a manufacturer, such as, "Please make us ten dozen 8/32 left-handed screws."

At a more fundamental level, molecules generally have definite reflection symmetries. A molecule can be invariant under a reflection—a singlet— such as H_2O, which looks the same in a mirror. Or a molecule can become a different one, having a mirror partner, when we reflect it in the mirror. A molecule that is the mirror image of another molecule is called a "stereoisomer." A stereoisomer pair contains left (levo-) and right (dextro-) forms that differ only by reflection in the mirror (like our left and right hands). The dextro-molecules are the mirror images of the levo-molecules, and vice versa. The dextro- (levo-) stereoisomers will have the *exact same* chemical properties when they are mixed into soup with other dextro- (levo-) stereoisomers. However, dextro- (levo-) isomers will have *different* chemical properties when they are mixed into soup with the mirror image levo- (dextro-) isomers.

Complex living organisms on Earth all evolved from simple primordial organisms. One compelling indication of this is associated with the handedness of the molecules out of which we are built. We share particular stereoisomers with all other species. When the primordial organisms formed, certain random events happened, causing one of them to use, for example, a levo-molecule for a particular function. The choice was random, like the flip of a coin, the incorporation by chance of a given stereoisomer into the organism happened by mutation. But once the choice was made, all of the subsequent progeny of this single organism inherited the same stereoisomer for the particular function. As the evolutionary sequence continued, all the life-forms that evolved through further mutations from this primordial organism also inherited the same random choice of stereoisomer for this function. The choice propagated down the long chain of evolution as the more advanced species came into being. So, we inherited this random coin flip from our primordial ancestors about three billion years ago, when the earliest living things were forming in the sludge of primitive Earth.

For example, most types of sugar molecules found in organisms on Earth are the right-handed kind, or dextro-sugars, though the mirror image levo-sugars can be produced commercially or in the laboratory. The digestive enzymes in our stomachs, however, evolved to digest only the dextro-sugars that we normally encounter on planet Earth—the molecules that come from the other living organisms that also evolved on planet Earth. These dextro-enzymes do not chemically interact (in the same way) with the levo-sugars, so the levo-sugars are not digested. However, the nerves in our taste buds savor the levo-sugars as though they were dextro-sugars. Thus levo-sugars can, perhaps, be used as sugar substitutes, since they taste sweet but are not metabolized, ultimately just excreted, and will not cause weight gain or decay of our teeth. Of course, one must always anticipate undesirable side effects.

It is fascinating to think that, if and when we go to another planet and are greeted by new life-forms, we may meet aliens that may look just like us but may actually have evolved with different stereochemistry. The aliens may, for example, digest only the levo-sugars, so their carrots, beets, and chocolate candies would contain only the levo-sugars. We would sit down with them to enjoy a wonderful alien meal that tastes just like good old home cooking but later find that we are still hungry and received no nutritional benefit from the alien food whatsoever. We might find ourselves having to survive on alien sugar substitutes.

It is also intriguing that we can, in principle, trace evolution, together with all other living things on Earth, uniquely to the primordial organisms on the planet through this stereochemistry. It could have easily been the other way—it's like the kickoff of the Super Bowl, all determined by the "flip of a coin," as one little proto-organism incorporated a metabolizer for dextro-sugar, and that random event propagated down the entire chain of life in animals and plants to all things alive today. Evolution is basically a form of complex physics and is a factual set of principles. One cannot understand a modern biological seminar, or a research program in genomics, without understanding evolution. We can, of course, alter things for society, for example, as some US school districts insist, by denying our children an education in evolutionary biology that equips them to live and compete in the modern world. This will simply aid in natural selection, making it easier for a smarter species to replace ours at a future date.

Moving back from biology to physics, we might ask, "Is the physical world, that is, are the laws of physics, invariant under the discrete symmetry of reflection?" Are the laws of physics in Looking-glass House really the same as ours?

PARITY SYMMETRY AND THE LAWS OF PHYSICS

Reflections are symmetries of dynamic physical processes and/or fundamental physical objects (like atoms) as well. For example, the laws of electrodynamics of charged particles, and of gravity, viewed in the mirror world, are the same in the nonmirror world. We call this grand reflection symmetry of physics *parity*. What does it mean for "laws of physics" to be invariant under reflections?

In essence, parity symmetry, both literally and mathematically, means viewing the world, including all of its physical processes, in a mirror, as if we were in Alice's Looking-glass House. We see physical objects in the mirror that move around, collide and interact, and obey a system of "laws of physics" very similar to the system of "laws of physics" that work on our side of the mirror.

Let's imagine that a cat named Tum (in our world; see fig. 16) jumps onto a slippery surface, the freshly waxed tabletop, and slides into a vase of flowers, which smashes onto the floor. Momentum, energy, and angular momentum are all conserved in this collision process (if, of course, we include the dissipation energy into sound and heat and the cost in energy of separating atomic bonds to break the vase, etc., when the flowers finally hit the floor, the total energy is indeed conserved). These are the laws of physics, all dictated by symmetry principles on our side of the mirror, including Noether's theorem.

In the Looking-glass House there is also a cat, which looks very much like Tum. We'll call him Mut (he is also a "him"). He, too, slides on a slippery table surface and collides with a vase of flowers and knocks them onto the floor. We can actually make detailed measurements of this collision to check that in Looking-glass House there is exact momentum, energy, and angular momentum conservation as well. As far as we can see, translational symmetry in space and time, rotational symmetry, and most of the other symmetries, all are valid in the mirror world as well. And so, we begin to believe that Looking-glass House, the world in the mirror, is subject to exactly the same laws of physics that our world is subject to.

We recall that a reflection is a discrete symmetry, because we either reflect or we don't—there is no 0.126 units of reflection; it's all or nothing. This symmetry is called, as we noted, *parity* symmetry of the laws of physics. In other words, the laws describing physical processes as viewed through a mirror should be the same as those describing the same physical processes on our side of the mirror, if *parity* is a good symmetry.

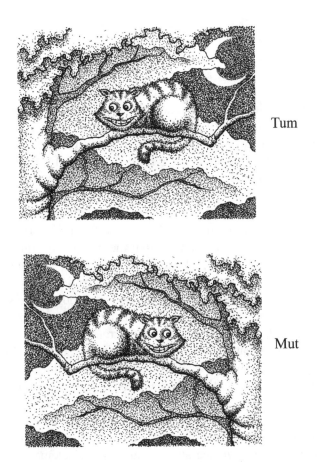

Tum

Mut

Figure 16. Tum, the cat, and his mirror image, Mut. (Illustrations by Shea Ferrell.)

Now, this raises an interesting and more precise question. The notion that Looking-glass House is governed by the same laws of physics as our world is a hypothesis. Is parity really a true symmetry of the laws of physics? How might we test this hypothesis to find out?

Suppose that you are given a movie, or a DVD, showing a physical system undergoing dynamic processes. For instance, this may be the collision of the cat Tum with a vase of flowers that falls on the floor. Or perhaps it is a simpler process, such as a collision of billiard balls on a billiard table. However, the movie may have been filmed with the camera *viewing the reflection of the system in a mirror*, as in figure 17. Let's assume that we had a really fine camera and an exceptionally clean and smooth mirror (no nicks or smudges). You were not permitted to see how the camera was set up and how it was viewing the billiard table. Is there any way you could tell that the physical process you were viewing was filmed through the mirror, as in image *a* of figure 17, or taken directly without viewing the reflection in a mirror, as in image *b*?

Now, this is a deep question, and we need to reduce it to simpler systems to appreciate that fact fully. Let's again consider the cat-vase collision, and suppose that I may have forgotten to tell you that Tum (a complex system) has a white spot on the right side of his face. Thus Tum is "tagged" with a distinguishing feature of right-handedness. Therefore when you view a movie of the cat-vase collision, you can look to see if the white spot is on the right or left side of the cat's face. If it's on the left side, then you know you are seeing Mut, the reflection of Tum, and you can tell that the image is viewed through the mirror. But this is not the issue. We have just seen that, like left-handed screws, we could in principle breed a new cat, named Ansel, who looks identical to Tum but who has the spot on the left side of his face. Then you couldn't be sure if the cat-vase collision is filmed on our side of the mirror with Ansel or in Alice's world with Mut! The spot becomes irrelevant.

So, we go to a greater level of simplicity and we look at billiard balls in collision. Can we now tell whether the film is taken through the mirror or not? Well? Yes or no? Ultimately, physicists prefer to go to the greatest level of simplicity and examine individual elementary particles in collision. Over the past century physicists have performed these experiments. At the highest level of resolution of our microscopes—the powerful particle accelerators—we can see atomic, nuclear, and elementary particle collisions, and most of the time they fail to reveal any difference between a given system and its mirror image. Indeed, up to the 1950s such observations had led physicists to think that, once we got down to truly elementary systems, systems that are not constructed from a complicated set of rules, such as a cat (involving natural selection and many stages of evolution in which handedness becomes imprinted), that we would always see pure left-right symmetric laws of nature. At this level we shouldn't be able to tell if the film is taken through the mirror and showing us the physics in Looking-glass House or taken directly and showing us our world. Thus parity was thought to be an exact symmetry of nature.

Nonetheless, scientists probed deeper into the sea of nature and methodically continued to test this idea. Are there any subtle properties of

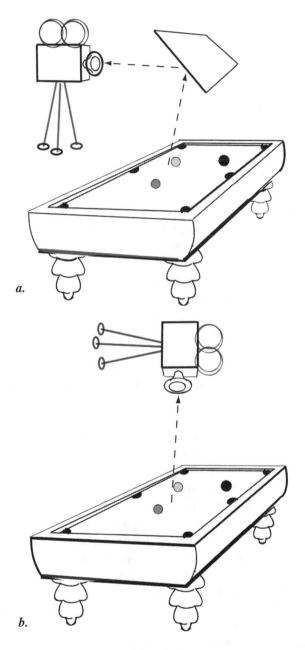

a.

b.

Figure 17. In *a*, a camera viewing a scene through a mirror, as it would appear in Looking-glass House; in *b*, the camera views the same scene directly, as it appears on our side of the mirror. (Illustration by CTH.)

elementary particles that are different in Alice's Looking-glass House from our world? Can we tell whether an imaginary film of atomic or subatomic processes is taken through the mirror or not?

THE OVERTHROW OF PARITY SYMMETRY

There is a particle called the pi-minus meson, or "pion" (pronounced PIE-on), denoted π^-. Today we know that the pion is not an elementary particle and that it is actually a composite object made of a "down quark" and an "anti–up quark," but for our present purposes we can think of it as elementary. The π^- decays within about a hundredth of a millionth of a second into two elementary particles, consisting of a *muon* (μ^-) and an electrically neutral *anti-neutrino* ($\bar{\nu}^0$), and we write this process as $\pi^- \rightarrow \mu^- + \bar{\nu}^0$.

The π^- is a "spin-zero" particle, meaning that it has zero intrinsic spin angular momentum. It can be considered a spherically symmetric, miniscule blob of matter, like a tiny billiard ball, which does not appear to change in any way if we rotate it. The muon, μ^-, and the antineutrino, $\bar{\nu}^0$, on the other hand, are like miniature spinning gyroscopes, each behaving like little pinpoints of matter with intrinsic spin angular momentum (they are called "spin-1/2" particles, but that detail need not concern us for now; the spins of elementary particles will be discussed in greater detail in chapter 10).

We know, according to Noether's theorem and rotational symmetry, that the conservation law of angular momentum must hold. This must be true even for *tiny elementary particles*, since rotational symmetry holds on all distance scales, and it is true in our world and the mirror world. Thus, when a π^- meson decays, the initial angular momentum is zero, therefore the sum of the final angular momenta of the μ^- and the $\bar{\nu}^0$ must be zero. The little gyroscopes of the produced muon and antineutrino must therefore be spinning in exactly the opposite directions so that the combined angular momentum adds to zero.

An extremely important experimental point, and the reason we can do this experiment at all, is that we can slow down and stop a speeding muon, and we can even measure its spin. Technically, the muon itself, in turn, decays (in a millionth of a second) into other particles, and the way in which its decay products unfurl tells us the muon's spin. Since the slowing down and stopping of the muon doesn't change its spin direction in space, we therefore know the exact direction of the muon's angular momentum (its spin) at the instant that it was produced from the decay of the pion.

So, we can set up an experiment to look in detail at the decay events of the π^-. We'll look for events where the muon comes out with its spin aligned *along the muon's direction of motion*. And we can also look for events in which the muon spin is *counteraligned to the muon's direction of motion*. When the spin is aligned in the direction of motion of a particle, we say the *helicity* of the particle is positive (+); when the spin is counteraligned to the direction of motion, we say the helicity is negative (–). Helicity is just a measure of handedness.

Since helicity is a form of handedness, like right- or left-handedness, the helicity of a particle is always *reversed* when viewed in a mirror (see fig. 18). To see this, recall how we consistently defined the angular momentum vector using the right-hand rule for something that is spinning. Consider, again, a toy gyroscope. For the gyroscope, we curl our right-hand fingers in the direction that the shaft is spinning, and our thumb defines the angular momentum vector direction (refer to fig. 10). This is a convention that we humans use, and *it must be used consistently for everything*—that is, we use the right-hand rule for muons and also for neutrinos. If we switch conventions somewhere in our thinking, we'll get the wrong answer. (We don't, for example, switch to the "left-hand rule" when we switch between muons and neutrinos. Also, since we don't know a priori if we are viewing a movie filmed through a mirror or not, we will *always* use the right-hand rule as we view spinning systems, *even for systems that we may be viewing through the mirror*. That is, we don't switch to a left-hand rule for a mirror image, because there is no way of knowing in advance if we are viewing the system through a mirror.)

Now consider a gyroscope that is spinning and moving in any direction, with its spin aligned in the direction of motion. Its mirror image may have the direction of motion reversed (if the gyroscope is moving into the mirror), but then the spin will not be reversed (using right-hand rule for the mirror image!). Or the direction of motion may be the same in the mirror, but then the spin direction reverses in the mirror, as in figure 18. So, we conclude that the helicity is always reversed in the mirror. As we said, helicity is a form of handedness, and handedness is always reversed in a mirror, just as your left hand becomes a right hand in the mirror, and vice versa. Also, you can think of the mirror image of a winding staircase, or of a screw, and you can see that helicity is reversed there as well (for

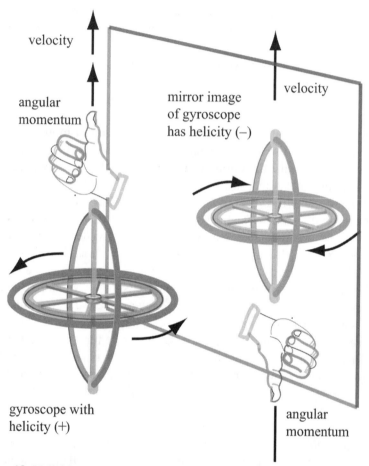

Figure 18. Helicities are always reversed in a mirror. Here a helicity (+) gyro-scope (angular momentum as defined by the right-hand rule points in the same direction as the velocity) has a mirror image that has helicity (−) (angular momentum as defined by the right-hand rule is reversed, whereas velocity is not). If the gyroscope axis together with velocity pointed into the mirror, then the angular momentum would not be reversed, but the velocity would, so the helicity would still be reversed, as it must be. (Illustration by CTH.)

a screw, this is the sense of the winding relative to the direction in which the shaft is shrinking to a sharp point).

One of the authors (LML), in the mid-1950s, measured the helicity of the outgoing (negatively charged) muon, produced in (negatively charged) pion decay, $\pi^- \rightarrow \mu^- + \bar{\nu}^0$. Let's try to guess what the answer should be for the outcome of this experiment. If parity was a good symmetry of the laws of physics, then both helicity (+) and helicity (−) muons should have occurred with equal probability (as we'll see later, quantum theory gives us only the probability of something happening, for many events, and it can't tell us exactly what will happen in any given event). That is, for many decay events we should get exactly 50-50 helicity (+) and helicity (−) muons coming out. This would have to be true by parity symmetry, because any given decay of the pion would have to produce a definite helicity for the muon, and the mirror image of any particular event would have the opposite value helicity. So any particular decay of the pion is different from its mirror image, and parity would require that things balance out over many, many decays. This is how old Aristotle might have reasoned it out.

In actuality, the result obtained by performing the experiment turns out to be shocking: the helicity of the muon produced in π^- decay is *always* negative; that is, we always see events as in figure 19*b*, never as in 19*a*!

So, why is this so shocking? It simply implies that if we ever "see" a movie film or a DVD recording of a π^- decay producing a helicity (+) muon, then we can loudly proclaim: "We are seeing an image of the process reflected in a mirror! Such a process can happen only in Alice's Looking-glass House. This never happens on our side of the mirror!" The mirror world is thus different from the world we inhabit in a fundamental way, at the level of the basic elementary particles and forces in nature.

Of course, the mirror world, with its helicity (+) muons coming from pion decay, is really a theoretical figment of the imagination and doesn't exist. In our world, the laws of physics contain forces and interactions that are not symmetric under parity, as in the particular "weak interactions" that are producing the decay of the π^-. Indeed, this is an example of parity symmetry violation that occurs throughout the weak interactions, which also produce numerous other effects in nature. Indeed, the very process that blows a Titanic star to smithereens in a supernova, the beta-decay process, $p^+ + e^- \rightarrow n^0 + \nu^0$ (proton plus electron converts to neutron plus neutrino), is a premier example of a weak interaction. As we've seen, the matter out of which we are composed, hence our very existence, depends upon these feeble forces in nature, and we now learn that these forces can distinguish our world from its mirror image!

Historically, as noted, until the mid-1950s physicists believed that parity was an exact symmetry of physics. The question of parity (designated as P) nonconservation in the weak interactions was first raised by

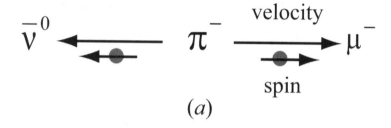

(a)

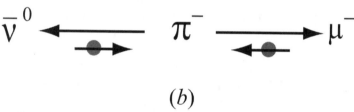

(b)

Figure 19. The helicities of particles produced from (negatively charged) pion decays, in the process $\pi^- \rightarrow \mu^- + \bar{\nu}^0$. In a the muon helicity is positive; in b muon helicity is negative. We always observe b in the laboratory, never a.

two young theorists, T. D. Lee and C. N. Yang, in 1956. Parity symmetry was practically considered to be a bread-and-butter, established fact in nature and had been used for decades in compiling data on nuclear and atomic physics. The conceptual breakthrough of Lee and Yang was the idea that the reflection symmetry—parity—could be perfectly respected in most of the interactions that physicists encountered, such as the strong force, which holds the atomic nucleus together, and the electromagnetic forces together with gravity. But Lee and Yang proposed that the weak force, with its particular form of beta-decay radioactivity, might not possess this mirror symmetry.[1]

In 1957, parity violation was discovered experimentally, by one of the authors (LML) and his collaborators, by using the pion decay technique we have just described.[2] Independently, the effect was seen by Madame Chien-Shiung Wu, using another, more complex technique. This was astounding news—the weak processes are not invariant under the parity (P) operation. King Parity was overthrown! This was a new and revolutionary idea—the forces of nature may have their own individual degrees of symmetry.

Madame Wu observed the radioactive disintegration of cobalt-60 (^{60}Co) at very low temperatures in a strong magnetic field. This experiment was a very challenging undertaking, requiring the heroic efforts of many groups with different areas of expertise. Cobalt-60 is a metal out of which ordinary electrons stream, coming from beta-decay processes within the material. Wu discovered that, in the strong magnetic field, the electrons were emitted *in the direction of* the magnetic field (this happens because the magnetic field, at low temperatures, aligns the spins of the nuclei in the cobalt, and the decay pattern is determined by the spin of the nucleus). However, her observation was enough to conclude that there was a violation of parity symmetry. The alignment of outgoing electron velocity with the magnetic field, it turns out, is the same as a helicity, and it would be reversed in a mirror.[3] If we saw a movie or DVD showing the electrons coming out of ^{60}Co decay counteraligned to the magnetic field, then, again, we could announce: "This is a mirror image of the real process and does not occur in our world."

TIME REVERSAL SYMMETRY

Again, let us think about viewing the laws of physics by watching a movie. Rather than viewing the film through a mirror, however, we now run the film backward through the projector. This is easy to do with a VHS or DVD player nowadays, by pressing the rewind or reverse button. We have all seen, with amusement, the pie fly off Uncle Bert's face, or brick towers uncollapsing and flying back into their original positions. Unlike the world viewed through the mirror, it seems very easy to proclaim that you're watching a film that is running backward through the projector.

However, we must again be careful to ask if this is truly a fundamental aspect of nature, or something that tags nature like the white spot on Tum's face. That is, if we saw a heap of bricks spontaneously form a neat tower of bricks, we could in fact say to a high degree of probability that the film was running backward through the projector. However, when applied to simpler systems, such as two billiard balls colliding on the table, it becomes harder to tell in which direction the film is progressing. The motion we see, forward or backward, as two billiard balls approach and bounce off of one another, recoiling off into different directions on the table, appears not much changed if we reverse the film. The forward-in-time collision seems to respect the same laws of motion as the backward-in-time collision. The laws of motion of simple systems are evidently the same, whether run forward or backward in time. But how can we ever really run the laws of physics back in time to test this hypothesis?

In physics we always pose, and solve, *if-then* problems. Let's consider the following elementary physics question (we will call it Q1): *If* a particle at time t_1 is located at x_1 traveling at a velocity V, *then* where will the particle be at time t_2? The answer is $x_2 = x_1 + V(t_2 - t_1)$.

Consider, now, a *time-reversed question* (Q2): "If at time t_1 the particle is located at x_2 and traveling with velocity $-V$ (velocities change sign when we reverse the direction of time, as you well know by running a DVD backward or seeing a car driving in reverse down the highway), then where will it be at time t_2?" Now the answer, by common sense, must be x_1. And indeed, we see upon a little rearranging that our previous formula gives us $x_1 = x_2 - V(t_2 - t_1)$.

This is indeed the correct answer for the time-reversed question, yet it came from the original problem's solution after a little rearranging of the math. The answer for the forward-in-time question evidently contains the answer for the backward-in-time question—we get both from one and the same physics equation! Our physical description of this system is the same if time is running forward as when time is running backward. In Q2, we set up initial conditions that were the *opposite* to those in Q1; that is, in Q2 we put the particle at the location x_2, where it ended up in Q1, and we reversed the direction of motion, replacing V with $-V$. We find that after an equivalent time interval, the particle in Q2 gets to location x_1, where it started in Q1. This shows that we can do time-reversed physics without actually reversing the flow of time. We need only reverse the directions of motions and swap the final destinations for the initial one. Put in its simplest form, a trip by train from New York to Philadelphia is the time-reversed version of the trip from Philadelphia to New York.

We often wonder why the more complex systems seem to indicate a preferred direction of time, or an *arrow of time*, whereas their elementary counterparts do not. Why does a tower of bricks fall down into a pile of bricks and dust, but no pile of bricks and dust ever falls up into a tower? Yet time reversing a billiard ball collision looks almost the same as a non–time-reversed collision.

This has to do with the *if-then* nature of physics questions. Anything we observe involves laws of motion but also particular initial conditions. If we initially have a container full of a gas, and we open the valve on the container, the gas will escape and fill a room. Here the initial condition is the fact that we start with a container full of compressed gas, which can be readily done with a compressor. The laws of motion for the escaping gas are then perfectly time-reversal invariant, but we never observe the time-reversed situation. That is, we never see a room full of gas collect itself spontaneously into the container. It is simply very unlikely to have an initial condition consisting of a bazillion gas molecules with their velocities and positions such that they will collect into the container. Such an initial condition would not violate the laws of physics, but it would be absurdly improbable. Seeing a bunch of billiard balls colliding into its "racked" configuration would look equally peculiar, while there is nothing peculiar about the break shot that scatters the balls apart from the racked configuration. The initial conditions are what make a situation look peculiar when we time reverse it.

In the physics of complex systems we can introduce a statistical concept, a measure of randomness, called *entropy*. In a quiet equilibrium, like hot onion soup sitting in a thermos bottle with no escaping heat, the entropy remains constant in time. However, in violent nonequilibrium processes, like shattering glass or explosions, the entropy always increases. Essentially, entropy, as a measure of randomness, will always increase when a very ordered initial condition leads to a very disordered final state through the normal laws of physics. The fact that entropy at best stays the same in equilibrium, or increases in all other processes, is called the *second law of thermodynamics*.

Now, this does not mean that complex, ordered systems cannot evolve and still respect the "second law." Indeed, by observation, they certainly can and do evolve. In a system such as a gas of water vapor cooling, the water will form condensed droplets of liquid, which have more statistical order (less randomness) than the original gas. With more cooling, the droplets form still more ordered, and less random, crystals of ice. In this process of cooling, energy has been allowed to flow out of the water vapor (perhaps as radiation, i.e., photons). As the departed energy scatters out into space, it occupies a more chaotic distribution (more entropy), whereas a small subsystem of cooled droplets is left behind (less entropy). The overall total entropy has increased, although a subsystem has formed with less entropy. If that subsystem contains a certain configuration of molecules, such as a nucleic acid (a building block of

DNA), then it may be able to make copies of itself by complex chemistry, yet still expending more energy out into space. Again, overall entropy increases, but we get a more and more complex subsystem left behind. And eventually, we can form a human being sitting there, wondering (and worrying) why time seems to flow in a particular direction for complex systems. The complex subsystem, (if) having been formed, (then) can evolve in a way by which it increases its own entropy: it can fall apart, rot, dissolve, or fade away.[4]

But is time reversal truly a fundamental symmetry of nature, and valid microscopically, for the elementary particles? Are all physical processes described by equations that equally well describe the time-reversed processes? Can we ask this question much as we did for parity and establish the answer by experiment? Indeed, we can. The answer is again shocking: the weak interactions, which violate parity, also violate time reversal invariance. However, to understand this, we must introduce *antimatter*.

TIME REVERSAL INVARIANCE AND ANTIMATTER

One of the most remarkable consequences of Einstein's special theory of relativity is that, when combined with the quantum theory, it predicts the existence of antimatter. The theoretical prediction of antimatter, by Paul Dirac in 1926, and its later experimental confirmation, is one of the most profound scientific results of the twentieth century. We'll see why antimatter must exist and study it in greater detail in chapter 10. It essentially comes from the discrete symmetries of space and time; hence, antimatter is intimately related to the symmetries of parity and time reversal of space and time. In fact, Richard P. Feynman gave a novel interpretation in 1949 of an antiparticle as a particle moving "backward in time."

Hence, for every species of elementary particle in nature there exists a corresponding species of antiparticle. For example, the electron, with its negative electric charge, has an antiparticle called the *positron*, with a corresponding positive electric charge. The positron has the exact same mass as the electron, and when it collides with an electron the two objects disappear, leaving behind photons to conserve the energy and momentum of the collision. At Fermilab we take antimatter for granted, as the Tevatron hurls protons in one direction and collides them head-on with antiprotons hurled in the opposite direction. Such collisions can make a pair of a new form of matter and antimatter, a top quark and an anti–top quark.

The existence of antimatter leads us to yet another discrete symmetry in nature: that of replacing all particles with antiparticles in any given reaction. This is called C, or "charge conjugation." This symmetry would imply that the laws of physics are exactly the same in the antiparticle world as they are in the particle world. For example, antihydrogen, consisting of an antiproton and an antielectron (positron), would have the same properties—for example, energy levels, sizes of the electron (positron) orbitals, decay rates, and spectrum—as does the ordinary hydrogen atom.

We have already noted that mirror symmetry, designated by P for parity, is not a valid symmetry when it comes to processes involving the weak forces. Furthermore, as we have seen, we can define yet another discrete symmetry operation, called "T," which reverses the flow of time; that is, we can replace t with $-t$ in all of our physics equations, swap initial conditions with final ones, and get the same, consistent results.

If C is a symmetry of physics, then an antiparticle must behave in every respect identically to its particle counterpart, provided we replace every particle with its antiparticle in any given process. But this makes no reference to the spins and momenta of the particles, which have to do with space and reflection (P) transformations. In the pion decay, $\pi^- \to \mu^- + \bar{\nu}^0$, the produced muon always has negative helicity. If we perform a C operation on this process we get the antiparticle process, $\pi^+ \to \mu^+ + \nu^0$, where all particles are now replaced by antiparticles, but the spins and momenta all stay the same as in the original process. Therefore, the helicity of the antimuon in the antiparticle process should still be negative.

In 1957, shortly after the overthrow of P, this was tested directly through experiment. When the experiment was performed, the helicity of the antimuon was *not* negative; rather, it was found to be positive. Therefore, C is also violated, together with P in weak interactions, such as the decay of pions and muons. Put into words, replacing particles with antiparticles everywhere in a given process is not a symmetry of the process, since it yields the opposite results (mirror images) for all the helicities for the particles in the process.

Naturally, there then arose the intriguing conjecture that, perhaps, if we reflect in a mirror, P, which reverses all the helicities, and we simultaneously change particle to antiparticle, C, then this combined symmetry may be exact in nature. The combined symmetry operation is called CP. Upon performing CP to the negatively charged *left-handed* (negative-helicity) muon, we get a positively charged *right-handed* (positive-helicity) antimuon. In the pion decay, $\pi^+ \rightarrow \mu^+ + \nu^0$, the muon observed in the lab indeed has positive helicity (is right-handed), so this turned out to be a symmetry of pion decay. Upon hearing this, physicists rejoiced! We seemed to have a deeper symmetry that connected space reflections with the identity of particle and antiparticle.

The joy was short-lived, however. In 1964, in a beautiful experiment, the "Fitch-Cronin experiment," involving some other interesting particles called neutral K-mesons (neutral K-mesons are composite objects, each containing a pair of strange and anti–down quarks, or down and anti–strange quarks), it was shown that CP is *not* conserved. That is, the physics of weak forces is not invariant under the combined operations C and P. The details of the origin of this breakdown of the symmetry CP has come to define a research frontier of physics for the past thirty years. We still do not know how this will play out, but we have since learned that if CP were indeed a perfect symmetry of nature, then our universe would be totally different, and our solar system, stars, galaxies, and we would probably not exist. Nor would you be reading this book. So it's a good thing for us that CP, as a symmetry of nature, is violated.[5]

The violation of the CP symmetry tells us that a particle and an antiparticle behave in slightly different ways. In fact, CP violation is welcome in the universe and is a necessary prerequisite to answering another enigmatic question: why does the universe apparently contain only matter, and no antimatter? If we go back to the initial instants of the big bang, when the universe was extremely hot (hotter than any energy scale ever probed in the lab), theory predicts that equal abundances of matter and antimatter were present. However, as the universe cooled, and with CP violation, some relic ultraheavy matter particles could have decayed slightly differently from their antiparticle counterparts. This asymmetry could have favored, at the end of the decay sequence, the production of a slight excess of the normal matter (e.g., hydrogen) over antimatter (antihydrogen). Then, as the universe cooled further, and much of the remaining matter and all of the antimatter annihilated each other, the slight excess of matter remained. This mismatched excess has since developed into us and everything we see in the universe. The open research problem is that, while we need CP violation to explain the fact that the universe contains matter and no antimatter, we don't think we have, as of yet, discovered the particular interactions that produce this effect. The CP violation effect, first seen in neutral K-mesons, now seen in other particle decays, remains an intriguing hint of much more to come. It is being studied aggressively around the world. The devil remains in the details.

PUTTING IT ALL TOGETHER: CPT

When we flip a "fair" coin, we have an equal probability of getting heads or tails. The sum of the probabilities of heads or tails is one—the sum of all probabilities that anything should happen must add to one, or else we are not able to talk meaningfully about probability. What would it mean for the probability of heads in a coin flip to be 2/3, if the probability of tails were also 2/3? Nonsense.

As we'll see later, quantum mechanics ultimately replaces the physics of Newton and Galileo, and it only makes probabilistic predictions for the outcomes of events in nature. Now, it turns out that it is a necessary condition in quantum mechanics that if we want the total probability of all possible outcomes for a given process to be conserved (i.e., the total probability for all possible outcomes must add to one), then the combined discrete operations of CPT, for any physical process, must indeed be an exact symmetry. That is, for any process in which we replace all particles by antiparticles (C), reflect them in a mirror (P), and run the camera backward in time (T), the outcome we predict should agree with the outcome that nature provides through laws of physics. If we combine C, P, and T, it turns out, at least at the present level of experimental sensitivity, that we do appear to have an exact symmetry of the world, called CPT, and the probabilistic interpretation of quantum mechanics works. There has been no experimental evidence of CPT violation, and many people consider it to be very unlikely. If CPT failed as a symmetry, then over time, probability would not be conserved, which undermines the notion of probability in quantum theory, and we would have to give it up. Nevertheless, we must ask, if the violation of CPT were very, very tiny, would we have noticed? It is, after all, an experimental question.

Suppose that while flipping a coin, a small black hole passed by and ate the coin. As long as we got to see the coin, the probabilities of heads and tails would add to one, but the possibility of a coin disappearing altogether into a black hole would have to be included. Once the coin crosses the event horizon of the black hole, it simply does not meaningfully exist in our universe anymore. Can we simply adjust our probabilistic interpretation to accommodate this outcome? Will we ever encounter negative probabilities? Do black holes eat probability in quantum theory, where they can instantaneously form and disappear in the vacuum itself? Is the CPT symmetry, or perhaps its violation (if any), connected to the other mysterious cosmic questions, such as the very origin of the universe? We have reached the frontier, and we don't know the answers to these questions—yet.

chapter 9

BROKEN SYMMETRY

Whence and what art thou, execrable shape?
—John Milton, *Paradise Lost*, bk. 2, line 681

Symmetries are often present in nature but hidden from view. This means that the symmetry is seemingly *broken* by the particular configuration of a system, the structure of a state of matter, or the state of the entire universe. The symmetry allows an equally likely but different configuration of the system. For example, the translational symmetry of our universe is hard to see, given that the Sun is nearby. The particular location of the Sun seems to imply a preferred center of the universe, or so the Aristotelians thought. But the placement of the Sun is a cosmic accident, a spontaneous choice, of one of an infinite number of equivalent places in the translationally invariant universe for a star to reside.

Indeed, there are many things we see in physics that exhibit no apparent symmetries. We have talked about the electron, the most basic electrically charged particle. We have seen that there also exists, however, the particle called the muon, which is in many ways identical to the electron but happens to be two hundred times heavier than the electron (and it quickly decays, in a millionth of a second, into an electron plus neu-

189

trinos through the weak interactions). We are very tempted to say that there must be a symmetry between the muon and the electron, and some corresponding transformation that relates an electron to a muon, and vice versa. Yet the big discrepancy in mass between these particles seems to get in the way—the particles really are quite different from the point of view of their masses. Could it be that a fundamental symmetry is really there but somehow hidden? Or is it the case that there simply is no meaningful symmetry between these two elementary particles? It is quite difficult to know for sure.

In certain systems that do not appear to have symmetry, however, a symmetry may definitely be present but hidden. Scientists can even understand how the apparent breakdown of the symmetry occurs, seeing definite relics of its presence that tell them what happened. This phenomenon is called *spontaneous symmetry breaking*. Indeed, it is likely that the universe itself began symmetrically poised in some grandiose way, as if a mathematical and symmetrical garden of Eden. The big bang explosion may have actually been a massive *symmetry-breaking event* that occurred in the subsequent instants of time. This grand symmetry breakdown may have given us the vast enormity of space and time, a near desolation of emptiness, through a process called "inflation." Getting back to Eden is the theoretical task of reconstructing that elegant initial symmetric state.

A Pencil on Its Tip

A little boy is seated at a table in the center of a circle. On the circumference of the circle are seated many little girls, with whom the boy would like to dance. How to choose? To choose one is to break the symmetry of the many equally attractive choices. This must be done in a fair and democratic way. There is no bottle to spin, but there is a sharpened pencil lying on the table.

The boy stands the sharpened pencil in an upright position, balanced upon its lead tip. The force of gravity, normally tipping the pencil over is, at the exact vertical position, completely balanced to zero. Gingerly he releases the pencil. Gravity pulls straight down, and the pencil in the exact vertical position doesn't lean in any preferred direction. The pencil hovers for a second or two, and suspense fills the room. The pencil momentarily seems to defy nature and reason and remains precariously perched on its pointed tip for almost an eternity.

Then, finally, whether it is a small vibration of the table due to an earthquake in Hong Kong or a tiny air current set in motion by a distant sneeze in Chicago, or a butterfly flapping its wings in a far-off rainforest in Costa Rica, or perhaps the gravitational rumbling of photon torpedoes in an interplanetary war in a galaxy far away, the pencil tilts, ever so slightly and imperceptibly, in some incalculable and apparently random direction, and it then topples over. The pencil comes to rest after a little bounce or two, its green eraser pointing in a direction that has evidently been chosen "by fate"—at random. The boy looks at the particular little girl, singled out by the direction of the pencil eraser, and he approaches her and asks her to dance. The decision has been made. The pencil has *broken a symmetry*—the symmetry of the

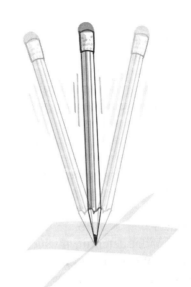

Figure 20. Pencil on its lead tip. The system (including gravity) has rotational symmetry about the vertical axis when the pencil is balanced on its tip, but this configuration is unstable. The pencil falls in a random direction, spontaneously breaking the symmetry. (Illustration by Shea Ferrell.)

many beautiful little girls seated about the circumference of the circle with whom the boy wanted to dance. The choice has been made at random, with great spontaneity. Hence, this is called spontaneous symmetry breaking.

When the pencil is perched in the vertical position, there is indeed a symmetry. Any of the potential dance partners is as likely as any other to be chosen. There is therefore a rotational symmetry about the vertical axis of the pencil shaft. This is a discrete symmetry, if there is a finite integer number of girls.[1] Physically, however, the gravitational force is completely balanced to zero when the pencil is perfectly vertical, and there is a continuous rotational symmetry about the axis of the pencil—any rotation about the vertical axis leaves the system, and its gravitational potential energy, unchanged.

However, this system is unstable. The symmetric state of the pencil perched upon its tip is a highly unnatural, "high-energy" configuration. The pencil eventually finds its way to a lower potential energy configuration by toppling over. Afterward, the pencil is pointing in some other direction in space. This can be any direction—the rotational symmetry implies the equivalence of all possible directions that the pencil can point. But a preferred direction has been randomly chosen. The symmetry of rotations about the vertical axis of the pencil is broken by the random direction selected by the pencil. Only one girl is selected to dance, though each was equally likely to have been picked.

There *are* apparently missing symmetries in the laws of physics. Why is the weak force weak, while electromagnetism is much stronger, and the strong force is stronger still? Why are there only three dimensions of space in which to rotate a flower vase, or in which to travel, and not more? What determines which symmetry holds and which symmetry is broken? Where has all the (possible) symmetry gone?

Or is there a more elegant way out? Can the laws of physics that we see controlling our universe, the laws of the elementary particles and their forces that ultimately drive the Titanic supernovas, the production of carbon and nitrogen, and the ultimate evolution of human beings, be governed by completely symmetric rules that are broken at random, spontaneously? These are excellent questions, and the answers appear to be, at least in part, yes. And the impact of symmetry breaking upon the universe at large appears to be dramatic.

MAGNETS

Magnets are counterintuitive and yet so much fun, as they embody a phenomenon that seems to defy naturalness. Ancient people thought they were of mystical origin or the work of a devil. The most common naturally occurring permanent magnets are composed of a mineral called magnetite, which is made up of a black iron oxide ore, Fe_3O_4. Shiny metallic magnets are often made of an alloy called alnico, containing aluminum, nickel, and cobalt. Even stronger magnets contain rare-earth elements, such as samarium and neodymium.

According to legend, magnetite was discovered by a Greek shepherd boy named Magnus, who noticed that certain minerals in rocks or stones attracted iron nails. The philosopher Lucretius later wrote that such stones had unusual powers, attracting and repelling each other. The Chinese may well have constructed the first compasses from magnetite many years earlier.[2]

In the thirteenth century it was noted in Europe that magnets always have two ends, or *poles*. One magnet's pole, let us say the "north" pole, is attracted to the other magnet's "south pole," and it is repelled by the other magnet's "north pole." The Europeans noted that, under delicate circumstances, one of the poles of a magnet naturally seeks out and points to the north star. The Europeans used compasses to navigate because the north-pointing end always indicated the direction of the north-pole star, even during daylight or cloud cover. Columbus used a compass when he crossed the Atlantic, and he noticed that the needle deviated slightly from the exact north (as defined by the stars), and the deviation changed during the course of his voyage. Sixteenth-century scientists realized that compass magnets point "north" because Earth itself is a gigantic magnet.[3]

Columbus's observations had thus indicated that the north magnetic pole and the north rotational pole of Earth are not the same! Over the history of the planet the north magnetic pole has migrated. Sometimes the north and south magnetic poles swap positions, as Earth's entire magnetic field reverses. Remarkably, we still do not have a precise theory as to why Earth is a large magnet and why it periodically changes over many centuries, sometimes dramatically reversing its direction altogether.

Refrigerator magnets are cheap, and we can buy them in many sizes and shapes, so they are ideal magnets with which to play and experiment. Often we get refrigerator magnets for free, as "promotionals" from our local neighborhood vendors. Some are flat and flexible, usually printed with a business card of a realtor or a pizza establishment on the flip side. Others are encased in plastic figurines or have decorative advertising attached. If we remove any plastic decorative advertising, such as a clown's head surrounded by various flavors of ice cream or the dentist's phone number attached to a plastic tooth, that contains the magnet, we are usually left with a black, ring-shaped object that snaps to a metal refrigerator or file cabinet door. Most everyone is tempted to play with these. We like to feel the force between a pair of them, in some positions, snapping together, and in other positions pushing apart. They almost seem alive. Some of us have, no doubt, contemplated *magnetic levitation*, perhaps even useful applications such as high-speed magnetic-levitation (maglev) trains.

A student friend of ours, Sherman, decided to do a science project. He obtained two of the refrigerator magnets and held them in a position at which they snapped together, thus binding to form a single, larger magnet. Sherman, carefully, then lit up his Bunsen burner and heated the refrigerator magnet pair to a fairly high temperature. As they became hot, the force of attraction weakened and, eventually, the two magnets fell apart. Holding the hot magnets with pliers, Sherman observed that the force mutually attracting the magnets had completely disappeared. The magnetism of the refrigerator magnets had been *destroyed by the heat!*

After a while, the magnets cooled back to room temperature. The magnetic force was still gone. However, Sherman had another, more powerful magnet, which was fully magnetized. He placed the cool, dead refrigerator magnets close to the stronger functioning one, even bringing them into physical contact. Voila! The refrigerator magnets became magnetized again. Upon closer examination, they were "recharged," with their magnetic fields pointing in the same direction as the larger magnet they had touched. The two refrigerator magnets, if placed together, would snap again into a pair.

Sherman then reheated the magnets, and again their magnetic force disappeared. He then placed the hot refrigerator magnets close to the stronger one, and still the force did not return. But now he let the refrigerator magnets cool while sitting near the stronger magnet. After cooling, the refrigerator magnets were again found to be "recharged." Their magnetic force had again returned.

All of this disappearing and reappearing magnetism does indeed seem mysterious, like some remarkable hocus-pocus from a Harry Potter book. There would appear to be some "essence" in refrigerator magnets that disappears or is driven off by heat but that can be coaxed back into the material. Does this essence flow from the larger magnet into the hot ones as they cool? Is it driven off, like a dangerous pox or vapor, as the magnets are heated? Perhaps this magnetic essence has special curative powers?

Indeed, even in our modern, scientifically enlightened era, this seemingly mystical behavior of magnets has led to belief in a kind of neovoodooism. "Magnet therapy" has become a billion-dollar business around the world. In particular, weak-field refrigerator magnets are sold with the promise that they will alleviate chronic pain and even cure horrible illnesses.[4] We scientists (who are trained to be skeptical) know of no physical or biological explanation for this kind of magnet therapy. At present, it cannot be concluded scientifically that weak magnetic-field therapy actually works, other than to produce a placebo effect. It is as likely to work as it is to be harmful—and most likely does nothing at all.

Robert L. Park observes that "therapy magnets" are basically the same as the flat, flexible refrigerator magnets that are used for business cards. He tested a pair of these from a "magnetic therapy kit" that cost about fifty dollars. The therapy magnets were so weak in magnetic strength that they failed to hold even ten sheets of paper pressed to a metal file cabinet. This means that the magnetic fields would barely penetrate human skin. "Not only do these magnets have no power to heal," he writes, "they don't even reach the injury. Magnets generally cost less than a visit to the doctor and they certainly do no harm. But magnet therapy can be dangerous if it leads people to forgo needed medical treatment."[5]

There are, nonetheless, organisms that are apparently sensitive to magnetic fields. Certain bacteria (called anaerobic bacteria) like to avoid oxygen and use Earth's magnetic field to sense direction. Inside the bacteria are grains of magnetite. The tiny bacteria that are floating in water and are too light in weight to feel gravity use the magnetic field to sense the direction "down" and to move away from the oxygenated surface and toward other living organisms at lower depths. Homing pigeons and even honeybees may be employing magnetite within their central nervous system to provide a navigational compass.

A magnet's field comes from the individual atoms out of which it is composed. The electrons orbiting the nucleus of the atom each possess intrinsic *spin* and *orbital angular momentum*, governed by the rules of quantum mechanics. The combination of the orbital and spin motion of a given electron is the *total angular momentum* of the electron. The spin and orbital motion of the electron produces a tiny electric current, which in turn generates a tiny magnetic field. Thus an atom itself can behave like a little magnet. Together the direction of the orbital and spin angular momentum of the electron determine the direction of the magnetic field for the atom. The atom ends up with its own "north" and "south" poles. Since different atoms have different arrangements of electrons in their orbits, they will therefore have different magnetic properties.

At very high temperatures, a ferromagnetic material such as magnetite, which contains iron (Fe), has a random alignment of its internal atomic magnets. The little atoms bounce around, changing their alignment, in the "heat bath" of crystal vibrations and the photon radiation of high temperatures. As the material cools, the atoms begin to settle down, and they begin to align themselves through their mutual forces with their neighboring atoms. Ferromagnetic materials develop many microscopic subunits called *magnetic domains*. Each domain contains billions of atoms, all aligned with their north poles pointing in the same direction.

As the magnetic material cools, without the presence of any other magnetic fields around, the different domains end up pointing in totally random directions. This is just a consequence of rotational symmetry. Within each magnetic domain, however, there is a preferred direction that has formed *spontaneously*, just like the pencil falling from its lead onto the table. The little magnet of one atom in the cooling environment influences its neighbor to point the same way, and these two in turn influence others to join. This extended alignment reaches only out a finite distance, until it meets the boundary of another magnetic domain. It is like the formation of a political party, with hundreds, then thousands, then millions of opinions, all aligned, which clash with another domain whose opinions are all aligned in another direction!

If we apply a strong magnetic field to a ferromagnet (or if the ferromagnet cools in the presence of a background magnetic field), we can force the alignment of all the domains. This causes the ferromagnet to *magnetize*. Once the applied, or background, magnetic field is removed, the domains remain aligned. With all of the individual magnetic domains aligned in one direction, there is now a strong magnetic field emanating from the material, which has thus become a magnet.

It may have occurred to you, thinking in terms of the poles of magnets, that there is something fishy about a ferromagnet. The alignment requires that the *north pole* of one atom become aligned with the *north pole* of its sideways neighbor (the neighbor on the left or right). However, as we have said, and, as a little experimentation with refrigerator magnets would quickly reveal, this is not what magnets like to do. The north (south) poles repel each other and are attracted to the south (north) poles. Therefore, to be a ferromagnet, the aligning force between atoms must be strongest between atoms in the vertical direction, whereby north pole of one atom is end-to-end with south pole of one above it. That this happens is a special, complex phenomenon and is ultimately associated with quantum mechanics. Ferromagnets are an exceptional case, and very few materials in nature are ferromagnetic. Some materials are *paramagnetic*, whereby the individual atoms act like magnets as in the ferromagnetic case, but they are weakly interacting with their neighbors or tend to counteralign with their neighbors, north pole to south pole, and so on, and produce no net magnetic field. These atoms can align themselves to an external magnetic field, but the alignment disappears as the external field is removed. On the other hand, almost all materials have a *diamagnetic* property. This means that the atoms (or molecules) may not themselves be little magnets but will develop into magnets and align themselves whenever we apply a sufficiently strong external magnetic field. Diamagnetism and paramagnetism are effects that usually are quite small, disappearing when the external magnetic field is removed.

Whenever ferromagnetic materials are heated above a certain temperature, known as the *Curie temperature* or *Curie point*, named for eminent nineteenth-century French physicist Marie Curie, they completely lose their magnetic alignment. The magnetic domains within the magnet are regenerated only when the material cools below the Curie point. This is called a *phase transition*. At high temperatures, with thermal radiation and lots of jiggling going on, the delicate magnetic interactions among neighbors become irrelevant. The material only knows about symmetry: the rotational symmetry of the world insists there is no preferred direction in space for the magnet to point, and the magnetic field thus disappears. The rotational symmetry of the magnet is restored at high temperatures.

SPONTANEOUS SYMMETRY BREAKING THROUGHOUT NATURE

Ferromagnetism is an exemplary form of spontaneous symmetry breaking in physics. At high temperatures the atomic spins are pointing randomly in space, and the system is statistically rotationally symmetric. When it is magnetized, at low temperatures, the spins are aligned in any of an infinite number of possible directions. Indeed, within the magnetic domains, like the pencil falling in a random direction, the spins become aligned, thus choosing that direction, spontaneously. The symmetry of rotational invariance at high temperatures appears to have been broken by a physical system that seems to know a preferred direction in space. But this happens by chance. It is a random selection of a direction in space, made by a single atomic pair, that becomes amplified as the system further cools, leading to a large effect seen at low temperatures.

Spontaneous symmetry breaking is a general phenomenon found throughout nature. In physical systems, it almost always happens because the symmetric configuration of the system has a higher energy than any of the nonsymmetric configurations. In the case of the pencil, when it is poised on its lead tip, it has a maximal amount of energy. This configuration has no net gravitational force acting upon the pencil to tip it over, yet it is unstable. A slight disturbance is enough to tilt the pencil, and then gravity begins to pull it farther away from its equilibrium. The pencil begins to fall over, in any direction, and it reduces its potential energy in so doing. Likewise, if we completely *disorder* all of the atoms in the ferromagnet, that is, we randomly point each spin in a different direction, we would then raise the energy of the system. When the system becomes very hot, then the energy is raised, and the system becomes completely symmetric again, and the magnetism disappears. At low temperatures, the system reduces its overall energy by aligning the spins of the magnets. The alignment begins initially in the small domains, but the individual domains do not become aligned until we apply the strong magnetic field. This is like smoothing out the wrinkles in a carpet, and the ferromagnet ends up in its true lowest-energy configuration.

In fact, most substances are chaotic and random configurations of atoms at high temperatures, when they are in the gaseous or liquid phase. As they cool and become solids, they typically form solids that have crystal lattices, which are regular and periodic arrays of atoms. Sodium chloride (ordinary table salt) forms an extremely regular cubic lattice, and this is reflected in the crystalline shape of pieces of salt examined under a microscope. Crystals, such as diamonds or quartz, can often be cleaved, or cut, in a way that exactly divides two adjacent planes of atoms, often giving rise to spectacular optical clarity. Ordinary ice, the solid state of water, is also a crystal. The crystalline state of matter, as it condenses from a gas or liquid into the solid state, spontaneously chooses the directions in space that will define the special planes and axes of the crystal. The rotational symmetry of the solid crystal is a discrete symmetry, and is actually a *smaller* symmetry than the full set of continuous rotations in space that defined the symmetry of the system when it was a liquid or gas, at a higher temperature. The rotational symmetry of space is therefore spontaneously broken to the smaller symmetry of the crystal lattice.

We can describe spontaneous symmetry breaking for many systems in terms of the celebrated "Mexican hat potential." Consider a large sombrero sitting on a flat table. The hat has a smoothly shaped top that descends down toward the broad circular brim. At the lowest point of the brim, there is a circular trough. Suppose we place a marble at the top of the hat. The marble perched there has a large gravitational potential energy. At the exact peak of the hat, the force of gravity is zero, but this is a precarious perch for the marble. The small jittering effects of thermal motion, or even quantum mechanics, are sufficient to give the marble a nudge. Once this occurs, the marble falls down the side of the hat and eventually ends up in the trough of the brim where the gravitational potential energy is minimized.

The marble's roll into the trough of the hat conserved energy, but we'll assume that much of the energy was dissipated into the usual energy wastelands. Once reaching the trough, the marble has now found a stable place of minimum potential energy in which it can reside. Figure 21 shows this for an arbitrary direction in which the marble has "chosen" to roll. In fact, the marble can end up at any point in the trough of the potential. All of these places have the *same* potential energy, because the original hat is rotationally symmetric about its axis.

There is therefore a remarkable consequence of the spontaneous breaking of a continuous symmetry. Because the system has chosen a direction in space for its alignment spontaneously—the marble has chosen one of the infinite number of equivalent points in the trough of the brim to come to rest—it actually costs us *no energy* to change this overall alignment. In other words, we can (on a completely frictionless table surface) rotate the hat. Likewise, we can rotate a ferromagnet, with no net expenditure of energy. That is, we can imagine doing the rotation very, very slowly, so no kinetic energy is involved, and the potential energy change from one position to another is zero. Also, if we consider the case of our toppled pencil, and the pencil is truly lying on a frictionless table, it can be made to rotate slowly. The boy can cleverly wait until the pencil points toward the next little girl and dance with her; with patience he can dance with every girl, as the pencil rotates around, and then repeat the process indefinitely, assuming all participants are sufficiently patient and agreeable to this arrangement.

Once the spins of the atoms of a ferromagnet are all aligned, there can be vibrations, or oscillations, in which an entire block of aligned atoms gently wave, like seaweed undulating in an ocean current. These are called *spin waves*. The main consequence of spontaneous symmetry breaking is that the longest wavelength spin wave corresponds to a rotation of the *entire system*. That is, if we simply rotate the system, at essentially no cost in energy to us (assume the system sits on a completely frictionless surface, or floats freely in space, and we do the rotation as slowly as we like), the motion of all the spins will resemble the gentle undulation of an infinitely long wave. Indeed, this means that the lowest-energy spin wave will actually have zero energy, corresponding to the uniform

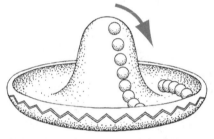

Figure 21. The "Mexican hat potential." (Illustration by Shea Ferrell.)

rotation of all of the atoms rotating the entire magnetic field in space with them. This long-wavelength, zero-energy spin wave, called a *zero-mode*, is one of the key indicators that a symmetry has been spontaneously broken in nature and one of the clues physicists look for to determine, in any given situation, if a symmetry is hidden from view.

COSMIC INFLATION

As we mentioned above, the enormous size of the universe itself is believed to be mostly a consequence of a phenomenon similar to spontaneous symmetry breaking. The key here is that spontaneous symmetry breaking relies upon the idea that the initial symmetric state of the universe, like a pencil standing on its tip, is actually unstable. Thus the state of maximal symmetry has maximal energy. In a sense, the system in the symmetric state is like an unstable bomb, ready to explode into an asymmetric state of much lower energy.

Let us imagine that a field exists in nature, permeating all of space, which we call the "inflaton" (IN-fluh-ton) field. This field, like an electric or magnetic field, can in principle take on any physical value in space and time. We postulate, however, that when the inflaton field has a value of zero, then the energy content of the inflaton field is large, like the potential energy of a marble perched precariously on top of the Mexican hat. This energy appears as energy of the vacuum itself, and it influences gravity and causes the universe to expand. When the inflaton field has a nonzero value, corresponding to the marble becoming located in the trough of the brim of the Mexican hat, the "broken-symmetry phase," then the energy of the vacuum is zero (or very nearly zero, as it is today). The expansion rate of the universe is therefore significantly less in the spontaneously broken phase, when the inflaton field has settled into the brim of the Mexican hat potential. Thus, with cosmic inflation, we have postulated that the universe began, with the inflaton field at the zero value, atop the Mexican hat, with an enormous vacuum energy (and, it turns out, negative vacuum pressure). The vacuum energy and pressure drove a rapid expansion of the universe, and the inflationary explosion ended when the inflaton field finally settled down into the trough of the Mexican hat potential, where the vacuum energy (and pressure) disappeared. This is called *inflation*. Vacuum energy and pressure have been converted to space and time in inflation.

This may seem to be a mad theorist's machination, but we already know that something like this must happen to break the symmetry of the observed forces in nature, particularly the electric and weak forces, and it also gives rise to the masses of elementary particles. This is called the "Higgs mechanism," and we'll return to it in detail in chapter 12. Cosmic inflation is an adaptation of the Higgs mechanism, due to particle theorist Alan Guth in the late 1970s. It is a mathematical depiction of the actual, and unknown, physics that caused the dramatic early expansion of the universe.

Indeed, there are many possible theoretical entities that could have provided a very large energy (and negative pressure) in the vacuum in the early phase of the universe. The theory of inflation does require a fairly long timescale over which the inflaton field, like the pencil perched on its lead tip, remains in the unstable, "perched" phase, and this makes the construction of realistic theories challenging. If the inflaton field quickly tumbled down to the state of lowest energy, where the vacuum energy is small or zero, then the net inflation of space would be small, and the universe would then be little more than a hiccup. Moreover, one can imagine a disaster akin to pencils that have tips that are glued in place and never fall—the analogue would be an inflaton field that got stuck in a dimple at the top of the Mexican hat potential and never rolled down into the brim. The analogue of this would be a universe that inflated forever, and all matter would be diluted away to a dismal nothingness of empty space and eternal time. Eternity is not always a good thing.

Remarkably, there is compelling astronomical evidence that something like inflation really happened. Inflation not only explains why the universe is so large but also why it appears to have the global symmetries on large scales of rotational and translational invariance. Indeed, the universe appears the same in all directions and appears to be the same in all places. We say that the universe is *isotropic* and *homogeneous*. All of this is actually difficult to explain through a big bang model of the universe without the process of inflation, usually leaving a bumpy universe with a different appearance in different directions of space.

Inflation unambiguously predicts that the overall energy density of the remaining matter in the universe today must be very close to a certain exact "critical" value that corresponds, through Einstein's equations, to an approximately infinite, or flat, universe. This is a consequence of the fact that the explosive expansion of the universe drove it to an almost infinitely large state. For the same reason, the universe would be almost homogeneous and isotropic, as confirmed by detailed observations of the cosmic microwave radiation. Finally, the observed fluctuations, bumps, and wiggles in the cosmic microwave background radiation—the radiation left over from the hot, initial phase of the big bang—are exactly as expected from the effects of a kind of quantum jiggling of the inflaton field during its roll down from the top of the Mexican hat potential.

Thus the enormity of our universe appears to be connected to the phenomenon of a spontaneously broken symmetry. Remarkably, however, we learn that our universe seems to have been largely an accident, like a pencil, or a tall tower, toppling over. We thus end up in a state of rubble, with much less symmetry than we started, making it very difficult to reconstruct the fossil record of fundamental physics in the earliest instants of creation. Our ultimate probe of the physics of the earliest instants of time are the most powerful particle accelerators that we can conceive of and build. Only these tools can reveal the original symmetries of nature in their fresh and unbroken state.

chapter 10

QUANTUM MECHANICS

The opposite of a correct statement is a false statement. But the opposite of a profound truth may well be another profound truth.
—Niels Bohr

Our accumulated understanding of the physical world through the beginning of the twentieth century is called *classical physics.* Based essentially on the formulation of Isaac Newton, it was tested repeatedly in thousands of experiments over a period of two hundred years, and it was always found to be correct. Newton's laws in the 1800s were subsequently supplemented by the laws of electricity and magnetism, established over many decades and neatly summarized in the mathematical formulation of James Clerk Maxwell.

Nonetheless, the data concerning the energy content of light and the idea of an atom did not fit the classical picture. Many questions began to accumulate. In about the year 1900, Max Planck, a German physicist, fretted over the color of light that is radiated by a piece of hot iron. At moderate temperatures, the iron would glow red, but when heated to much higher temperatures it would turn blue-white. Yet his detailed calculations, using Maxwell-Newtonian physics, predicted that it should glow blue at all temperatures, dim blue at low temperatures and bright blue at high temperatures. Planck recognized there was a serious problem with the Maxwellian theory of light—it simply didn't get the energy content of a beam of light right. Solving this problem launched the revolution and led to the new physics we call *quantum mechanics*.

Quantum mechanics evolved over a period of about thirty years, from about 1900 through 1930. At that point, the new theory was a stunning success, completely redefining how we think about the physical world. This was no mere academic exercise in existential philosophy—today, quantum mechanics and the understanding it brings of the electron, the atom, and light account for a major fraction of the US gross domestic product. It is the basis of all the known laws of physics and is the essential key to unlocking the deeper mysteries about matter and the universe.

Quantum-mechanical effects appear in physical systems that are *exceedingly small*. A *small system* means very tiny objects with very tiny amounts of energy, moving around over very short time intervals. Quantum effects show up dramatically once we arrive at length scales the size of the atom, about one ten-thousandth of a millionth (10^{-10}) of a meter. In fact, we simply cannot understand an atom without quantum mechanics.

This is not to say that nature itself suddenly "switches off" classical mechanics and "switches on" quantum mechanics when we enter this new submicroscopic realm. Quantum mechanics is always valid and always holds true at all scales of nature. Rather, quantum effects gradually become more and more pronounced as we descend into the world of atoms. Quantum mechanics is the ultimate set of rules, as far as we know, that governs how nature works. It is also very bizarre. It has been said that no one truly "understands" quantum mechanics—scientists simply get used to dealing with its weird rules.

We sometimes try to describe quantum effects in the following way. In the *macroscopic world* of many, many atoms, making up large objects like planets and people, slowly moving about like elephants, the effects of quantum mechanics become almost imperceptible. The classical descriptions derived from Newton take over as a kind of "average" effect. As a metaphor, consider a national census poll showing that the average household in the United States has exactly 2.27 children. The poll is precise, with a statistical error of perhaps ±0.01. This would be our analogue of a system that is described by the laws of classical physics—Newton's equations would predict that the average household should have any value of a *continuous number* of children, and the experiment thus reveals the number is 2.27 ± 0.01. Nonetheless, at the level of individual families—the "microscopic" level—there are *no households* with 2.27 children! (You are not surprised, are you?) All family households are actually found to be *quantized*, having a *discrete* number of children: 0, or 1, or 2, or 3, and so on, and only the *average* over many households yields the "classical," noninteger result of 2.27.

The larger a physical system becomes, the more it generally appears to behave as the average of its constituents—the more classical it becomes. However, even this simple example cannot begin to capture the bizarre essence of quantum mechanics. The effects of quantum mechanics, as we hope to show, are far spookier than mere statistical averaging.

Quantum effects do sometimes show up dramatically on extremely large distance scales, in macroscopic systems. Things like neutron stars, supernovas, and modern household gadgets employing lasers (e.g., CD and DVD players), as well as the remarkable phenomenon of superconductivity (electric current flowing at zero resistance) are direct quantum effects. For that matter, all of chemistry, and hence biology, is sculpted by quantum mechanics. The very structure and distribution of matter throughout the universe appears to be a consequence of quantum mechanics. We live in a quantum-mechanical world.

IS LIGHT A PARTICLE OR A WAVE?

For centuries, perhaps beginning with contentious arguments between Isaac Newton and Robert Hooke, scientists debated the issue of whether light was a *wave* or a *particle*. Light casts shadows, moves unobstructed in straight lines, and stops as it hits an object—similar to what one might expect from a stream of tiny bullets.

Yet light also undergoes *diffraction* and *interference*, causing undulating patterns, as it tries to pass a sharp edge or through a narrow slit. These patterns are characteristic of a wave, such as a water wave, as it passes by an object that disturbs the surface. The question remained until the early twentieth century—is light a particle or a wave?

In the nineteenth century we learned from James Clerk Maxwell and his theory of electromagnetism that light is a traveling wave of electric and magnetic fields. Therefore many physicists believed the puzzle to have been solved—light was conclusively thought to be a wave, traveling at the speed of light, *c*, transporting energy from the *source* of the light to the *receiver*. The theory established that light is produced by rapidly acceler-ating electric charges and is absorbed as it causes distant electric charges to accelerate. These things were tested experimentally, and even the first radio transmissions were established in the late nineteenth century using this successful theory. A light source can be anything that sufficiently shakes, wiggles, or jostles electrons, thus causing them to accelerate.

We can draw upon the example of a campfire to understand the phe-nomenon of light. The electrons in the atoms in a hot campfire are "ther-mally agitated," colliding with one another and with the light of the campfire itself, emitting and absorbing waves of light as they accelerate when they recoil in these collisions. The light travels away from the source, and some of it eventually enters our eyeballs, where it jostles and wiggles electric charges in the receptor cells within the retina of the eye. Here the light wave is absorbed and deposits its energy. By virtue of this electronic wiggling, a chain of chemical reactions is initiated that pro-duces a nerve impulse that travels into the visual system of our brains. We then enter the realm of consciousness and perceive the soothing scene of a campfire on a cool summer night.

Radio waves are also a form of light, outside of the range of sensi-tivity of our eyes, and therefore invisible. Indeed, the antenna that trans-mits a radio wave is a long wire in which an alternating current of elec-tricity—accelerating electrons—is created, which emits the radio wave. The radio receiver likewise has an antenna in which the incoming radio wave accelerates electrons, producing an electrical current that can be amplified by the receiver circuitry to produce a serene ballad by Norah Jones or a haunting symphony by Gorecki. Maxwell's theory is still used to this day to design antennas. Moreover, it formed the exclusive basis of most of electronics through the mid-twentieth century.

What is a wave? Consider a long, traveling wave as it moves through space. A traveling wave is sometimes called a *wave train*, with many sequential crests and troughs of the train as it traverses space. Such a wave is described by three quantities: its *frequency*, its *wavelength*, and its *amplitude*. The wavelength is the distance between two neighboring troughs or crests of the wave. The frequency is the number of times per second that the wave undulates up and down through complete cycles at any fixed point in space.

If we think of the wave as a long freight train, then its wavelength is the length of one of its boxcars. Its frequency is the number of boxcars per second passing in front of us as we patiently wait for the train to pass.

The speed of the traveling wave is therefore the length of a boxcar divided by the time it takes to pass, or, mathematically put, the speed of the wave equals the wavelength times the frequency. Thus, knowing the speed, the wavelength and frequency are *inversely related*; that is, the wavelength equals the speed of the wave divided by the frequency, and the frequency equals the speed of the wave divided by the wavelength.

The *amplitude* of the wave is the height of the crests, or the depth of the troughs, measured from the average. That is, the distance from the top of a crest to the bottom of a trough is twice the amplitude of the wave, and it can be thought of as the height of the boxcars. For an electromagnetic wave, the amplitude is the strength of the electric field in the wave. For a water wave, twice the amplitude is the distance that a boat is lifted from the trough to the crest as the wave passes by. Figure 22 says it all.[1]

The *color* of a visible-light wave was understood in the nineteenth-century Maxwellian theory of electromagnetism to be determined by the wavelength (and, inversely, the frequency). If we take the frequency to be small, we correspondingly find that the wavelength becomes large. Visible light of a greater wavelength is red, whereas that of a lesser wavelength is blue.

Visible red light has a wavelength of about $6.5 \times 10^{-5} = 0.000065$ meters. The longer the wavelength, the deeper shade of red the color becomes, until it fades from our eyes' sensitivity at a wavelength of about $7 \times 10^{-5} = 0.00007$ meters. Taking the wavelength still larger, we have *infrared light*, which we can feel as gentle heat but cannot see with our eyes. As the wavelength becomes still larger, we enter the realm of microwaves, and still longer, we have radio waves.

On the other hand, as we take the wavelength to be shorter than about $4.5 \times 10^{-5} = 0.000045$ meters, light becomes blue. At shorter wavelengths (i.e., higher frequency) it becomes deep violet blue, and then with still shorter wavelengths, it fades from visibility, at about $4 \times 10^{-5} = 0.00004$ meters. At still shorter wavelengths light becomes *ultraviolet*, later becoming x-rays and eventually gamma rays, at much shorter wavelengths.

The key problem with the classical theory of light had to do with its energy content. The classical electromagnetic theory of light predicted that the energy in the wave depended solely upon its *amplitude*. The energy of an electromagnetic wave was thus predicted to be *independent of its wavelength*, or color. Red light and blue light of the same amplitude, or intensity, therefore would have had exactly the same energy content.

You don't have to contemplate a block of hot iron to see the problem.

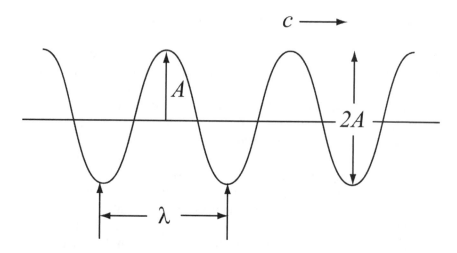

Figure 22. A wave train, or traveling wave. The wave moves to the right at a speed of c and has a wavelength λ (length of one full cycle, from trough to trough, or crest to crest). As a stationary observer watching the wave travel past, we would see a frequency of c/λ crests, or troughs, passing by per second. The amplitude is the height of a crest above mean (i.e., the average).

This concept of the energy of an electromagnetic wave raises immediate problems that are readily visible to our eyes. For example, when the coals of our summer campfire are cooling off, they glow red. The temperature of the coals is just a measure of the *average energy* of all the microscopic parts of the hot coals; all the various states of motion and vibration of the atoms, and electrons, that have approximately the same energy (equal to the particular value of the temperature) should be excited with equal probability. A very hot fire can therefore excite high-energy states of motion, and the atoms will move about with greater kinetic energies, and thus correspondingly higher-energy light waves should be emitted as radiation. A block of ice, on the other hand, has a low temperature; hence, only very low-energy states of motion and vibration of the atoms can be excited. The block of ice can only radiate small amounts of very low-energy light.

So why is no blue light emitted by the cooling campfire coals? Blue light, after all, has the same energy as red light, according to the classical theory of electromagnetism. As the coals cool, they glow fainter and fainter, and redder and redder, finally fading away into the warm, invisible, infrared light. If the classical theory was true, the fading coals should have emitted *just* as much blue light as red light at *all temperatures* (and also give off lots of x-rays and gamma rays as well!). When Max Planck did the calculation, he found that dying campfires should actually appear to glow blue. Technically, this should occur because many more short-wavelength blue waves than long-wavelength red waves could squeeze into the volume of space surrounding the hot coals. But it didn't jibe with reality, and Planck was right to think that something was quite amiss with the classical theory of electromagnetism.[2]

Planck proposed a radical remedy to this conundrum. He suggested that light contains component elements, or particles, that somehow move in a wavelike manner. These little elements of a light wave were dubbed *quanta*, or *photons*, as we refer to them nowadays. Planck proposed that each photon has an energy that increases in proportion to the frequency of the light wave, represented by the equation $E = hf$, where E stands for the energy of a photon, h for a fundamental constant, and f for the frequency. The intensity of the light is just a measure of the *total number of photons* in the light wave. Any light wave must contain a certain number of photons and will have a total energy $E_{total} = Nhf$, or the total energy of the wave equals the number of photons (N) times the constant h times the frequency of each photon (f).

The more intense a light wave, the more photons are present. But now the energy of each photon depends upon its *frequency*, with blue photons having a greater frequency, hence a larger energy, than red photons. This hypothesis "explains" why blue light is harder to excite than red light in the dying campfire coals. The lower-energy red photons can be more easily excited at lower temperatures, whereas the more energetic blue photons are harder to excite at lower temperatures.

Initially, many people thought that Planck's idea applied only to thermal situations. But the problems with the energy content of light became even more intolerable as more experiments in other areas were conducted. One of the most troubling of these was the *photoelectric effect*. Physicists found that they could readily cause electrons to pop out of certain metals by merely shining light on them. This has become the basis of modern television cameras, and digital cameras, which convert light into electrical signals. However, the photoelectric effect further defied the classical theory of electromagnetic radiation.

It was found that, for a given metal, red light did not eject electrons, whereas blue light did. In fact, the more bluish the light, the more energy was found to be present in the ejected electrons. There should have been no difference, according to the classical theory, since light energy did not depend upon the wavelength, or color, of light. No matter how bright a red light shone on metal, no electrons were ejected. With dim blue light, however, a few electrons were ejected from the surface of the metal. With bright blue light many electrons were ejected from the metal.

Einstein, in his *annus mirabilis*, his great year of 1905 (when he wrote about five Nobel Prize–caliber papers, including the introduction of his special theory of relativity), realized that Planck's new idea neatly explained the photoelectric effect. The electrons in the metal experience collisions with individual photons. If a single photon has insufficient energy to knock an electron out of the metal, then no matter how many photons are present, we'll see no ejected electrons. Thus, even if we have very intense red light—that is, very many photons but with each photon having lower energy—we expect no ejected electrons.

On the other hand, if the light is blue and each photon thus has larger energy, then each photon that strikes an electron will eject the electron from the metal. With dim blue light, containing few photons, we'll only get a few ejected electrons, but with intense blue light, containing many photons, we get many ejected electrons. We can actually count the photons by counting the ejected electrons! Einstein ultimately received a Nobel Prize, not for special relativity or general relativity but for explaining the photoelectric effect.

As we noted above, Max Planck, through this analysis of the color of thermally produced light, had invented the "magic" constant that defines quantum mechanics. This is called *Planck's constant* and is denoted by the letter h.[3] Planck's constant tells us the energy in a photon for any given frequency. In fact, Planck's constant and the speed of light are probably, to this day, the two most important physical constants in nature known to us (Newton's gravitational constant, which sets the fundamental scale of mass in string theory, is considered to be of equal and fundamental importance by many theorists). The constant h determines the crossover to what we mean by "small" in physics, or the onset of quantum behavior (much like the speed of light determines when the effects of special relativity will replace those of Newtonian mechanics). If the motion of a physical system involves energies and timescales (or momentum and distance scales) that, when multiplied together, yield a result comparable to or less than h, then we are in the *quantum realm*.

The physical value of h is precisely measured to be 6.626068×10^{-34} kilogram-meters squared per second. This is a very small number, characterizing the tiny distance, time, and energy or momentum scales that define the quantum realm.

QUANTUM THEORY BECOMES EVER MORE CURIOUS

Thus the embryonic quantum theory was taking shape. So far it seemed to pertain only to the behavior of light, where the paradox of wave-particle behavior appeared to be most pronounced. Perhaps, however, this also happened in other situations involving "periodic," or oscillating, behavior, similar to waves.

We now know that everything is made of atoms. A mosquito's eyelash contains perhaps a billion of them. A new picture of the structure of the atom was beginning to form. Some aspects of the atomic structure were understood by the time of Planck and Einstein, more or less, from experiment. From a series of key experiments performed by Ernest Rutherford from 1906 through 1911 at Cambridge University, it was known that there is a tiny, hard core inside of an atom, called the *nucleus*, wherein resides 99.98 percent of the atom's mass.[4] The nucleus had a large positive electric charge, and it was more or less realized that electrons, discovered in 1898 by J. J. Thomson, with their negative electric charges, are somehow orbiting the nucleus. It was becoming clear that the atom is a solar system–like object, with the nucleus at the center, like the Sun, and the electrons orbiting, like planets, around it. But again there were severe theoretical problems with the picture that emerged from the Maxwell theory of electromagnetism and energy.

When placed in orbits, electrons must accelerate—in fact, we have seen that all circular motion is accelerated motion, since the velocity is continuously changing direction with time. And, according to Maxwell's theory of electromagnetism, accelerated charges must give off electromagnetic radiation—that is, light. Estimates showed that *all of the electron's orbital energy* would be instantaneously radiated away into electromagnetic waves. By Maxwell's theory, then, the orbits of the electrons, and the atom itself, would thus collapse. Such collapsed atoms would be chemically dead and useless. Once again, nothing about the energy of electrons, atoms, or nuclei seemed to make sense in the classical theory.

Moreover, scientists of the nineteenth century knew that atoms do emit light but only in distinct spectral lines having definite colors, or distinct (*quantized*) values of the wavelength (or frequency). It seemed as if only certain special electron orbits exist in the atom, the electron hopping to and fro between these orbits as it emitted or absorbed light. A Kepler-like picture of the orbits would have predicted a continuous spectrum of radiated light, since there is a continuous set of possible Keplerian orbits.

It was as if the world of the atom was "digital" and far from the continuously varying world of Newtonian physics.

In 1911 Niels Bohr was a young researcher with Ernest Rutherford at Cambridge University, and he believed that the quantum theory would rescue the atom, as it did for the photoelectric effect and for the color of hot iron. Bohr's idea was that the electron's orbits were indeed like those of particles—planets going around the Sun—but they were simultaneously and paradoxically like waves. How then to apply the concepts of the new quantum theory? Bohr focused upon the simplest atom, hydrogen, consisting of a single electron orbiting a single proton as the nucleus.

Bohr realized, in 1911, that if the motion of an electron is like that of a wave, then the distance that it travels, through a complete cycle of its orbit (the circumference of the orbit), must be a distinct number of *quantum wavelengths* of the electron's motion viewed as a wave. This, Bohr argued, is related, through Planck's constant, to the magnitude of the electron's momentum in its orbit. That is, the momentum of an electron equals Planck's constant, h, divided by the quantum wavelength. The key to the atom is that this must match the orbital circumference, which equals an integer number times the wavelength. Therefore, the electron's momentum can only take on certain special values that are related to the size of its orbit. This is how a musical instrument works— only certain distinct wavelengths of sound can be produced from a brass tube of a given size, or a drum head of a given diameter, or a string of a given length.

Putting this together, Bohr immediately discovered that only a discrete set of particular allowed orbits, with distinct energies, were allowed for the electron's motion in an atom. The electron can occupy only one of these orbits at a time, but it can jump between them as it emits or absorbs light. Bohr's predicted energies of the emitted photons agreed precisely with the observed light emitted from hydrogen gas when it is extremely heated (usually by applying a spark of electricity in a tube containing hydrogen gas). The rudimentary properties of the hydrogen atom were now emerging, but many details still remained puzzling. It was still unclear what quantum mechanics really was. What are its true and universal rules? Does it apply only to electrons in orbits, and light? Or is it more general?

Finally came the liberating realization that *all particles in nature, under all circumstances*, always behave as quantum particle-waves.

Louis de Broglie, as a young graduate student in 1924, proposed that the electron, like light, is a quantum particle-wave under all circumstances, and it should therefore be possible to observe diffraction and interference patterns in the wavelike motion of untrapped electrons, just like those seen in light. He wrote down the relevant equations in a brief, three-page doctoral dissertation at the Sorbonne in Paris. The key was already contained in Bohr's idea that the momentum of the particle equals h divided by the wavelength; hence, given the momentum of a particle, we can compute its wavelength. But the thinking was now outside the box—it applied to *any* particle, anywhere, at anytime, not just those moving in circular orbits!

The distinguished faculty of the Sorbonne was unable to comprehend de Broglie's dissertation and was ready to dismiss the thesis altogether and flunk him. Fortunately, someone sent a copy of it to Albert Einstein with a request for a second opinion. Einstein replied that the young man in question deserved a Nobel Prize more than a doctorate. De Broglie was not flunked after all.

The wave properties of freely moving electrons were indeed observed in 1927 in a famous experiment at Bell Labs by Joseph Davisson and Lester Germer. Electrons were seen to undergo diffractive interference, like light waves, as they bounced off the surface of a crystalline metal. This was an astonishing development. No one ever questioned that electrons were anything but particles, yet they, too, also behaved like waves. After that, de Broglie was in fact awarded the Nobel Prize for physics in 1929. The pieces of the quantum puzzle were being put together into a completely new reality of nature.

THE UNCERTAINTY PRINCIPLE

Now, there appears yet another bizarre phenomenon in the world of quantum mechanics. The rules of quantum mechanics were being formulated at the University of Göttingen, where Emmy Noether had proven her theorem and now pursued her mathematical research in abstract algebra. There a brilliant theorist named Werner Heisenberg was also developing the system of mathematics that precisely defines quantum mechanics. It became clear to Heisenberg that the new quantum rules forced undeniable uncertainty into physics.

To understand this, let's do a thought experiment (or "gedankenexperiment"). We'll pretend that Planck's constant is not the tiny quantity we quoted earlier, but rather an enormous number. Let us suppose that it has a value of one unit, but in a system of units in which the unit of mass is the mass of our car, the unit of distance is the length of the state of Nebraska, and the unit of time is one hour. What would we experience on a road trip from Chicago, Illinois, to Aspen, Colorado, as we crossed Nebraska?

Suppose we measure the velocity of our car as it passes a highway mile marker on Interstate 80 somewhere in the middle of Nebraska. Let's say our speedometer says exactly 60 miles per hour. We check it several times and even set the cruise control at that number. There is no error in this speedometer because we are driving a fine German import, and we took out a big loan to buy it. This is an accurate speedometer!

Now we look out the window for the nearest highway marker and it says "Mile 186," so we are some distance from Omaha, going west. We have now measured our position precisely along the length of Interstate 80. At the same instant, we look back at the speedometer and remeasure our velocity. To our consternation, it says that we are now traveling 250 miles per hour!

We double-check the speedometer and reset the cruise control. Then we look back out the window, and we measure our position again by observing the next highway marker, and it says "Mile 30." We have actually gone backward and are now closer to Omaha, even though we were going west and passed Omaha two hours ago! We glance back at the speedometer and it now says 12 miles per hour! This is uncanny—we had better stop for gas, and an aspirin. But when we look at the next upcoming exit ramp, the mile marker says mile 320, and we are now at the west end of the state, near Ogallala!

As we try to come to a complete stop at the exact position of the gas pump, we find that we cannot stop—we, and our speedometer, are going crazy, traveling 50 miles per hour, then 400 miles per hour, then 136 miles per hour. We finally slam on the brakes, and come to absolute zero velocity, but when we look out the window, we see only a blur—Omaha here, Kearny there, the Rockies of Colorado over there, and Chicago here. We are at rest at exactly zero velocity, and yet we are everywhere in space at once! And, again, when we finally make sure we are located precisely at the gas pump, we find that we have all possible random velocities at once! We cannot seem to be *exactly somewhere* with an *exact velocity* (or momentum, which, you recall, is our velocity times our mass) at the same time.

Every time we precisely measure our velocity (or momentum), by looking at the speedometer and making sure it is reading a fixed speed, we find that we have randomly affected our position in space. And every time we measure precisely our position in space by noting a nearby mile marker, we have randomly changed our momentum (velocity).

This is a pretty bizarre nightmare, something out of the classic Rod Serling *Twilight Zone* series, yet it would be the true reality if Planck's constant were such an enormous number. We would then be a quantum particle-wave ourselves. Thankfully, Planck's constant is a small number, and only tiny particles, like electrons, must suffer this fate.

In quantum mechanics this is *reality*. We may know exactly the momentum of an electron, but it is then simultaneously at all places at once. Or else, we may know exactly where the electron is, but then we find it has all possible momenta (velocities) at once. We can "sort of" localize the electron in a region of space and "sort of" know its momentum, simultaneously balancing the uncertainty in the two quantities. The smaller the region into which we entrap an electron, however, the more uncertain becomes its momentum. We thus find that it takes an enormous force to confine the electron into smaller and smaller volumes because of the larger and larger uncertain swings in momentum.

Indeed, an atom manages to strike a balance through the electromagnetic force, localizing an electron in space into an *orbital* and yet providing enough force to keep it there as its momentum randomly jostles about. This is why atoms don't just collapse in quantum mechanics (while, you recall, they would have collapsed in Newtonian physics, where Planck's constant is zero, $h = 0$). The orbitals of electrons in atoms therefore don't look anything like Kepler's orbiting planets about the Sun. They are fuzzy things, trapped waves, the electron never having a definite position and momentum at the same time. We thus often refer to the motion of the electrons about the nucleus of an atom as the "electron cloud."

Stated more precisely, the uncertainty in the momentum times the uncertainty in the position will always be larger than Planck's constant divided by 2π. This effect is known as the *Heisenberg uncertainty principle*.[5] We emphasize that this effect is real, and it cannot be eliminated or reduced by better instrument making or finer tuning of a device. The more precisely we determine an object's momentum (velocity), the less precisely we can know its position, and vice versa. The product of the uncertainty in momentum with the uncertainty in position is—Planck's constant.

The inverse relation between momentum and wavelength (or uncertainty in momentum and uncertainty in position) has practical implications. To study something very small we need to make a probe that is smaller than the thing we want to study. So, too, in microscopy, the wavelength of the probe that we use must be smaller than the object we wish to study.[6] Since visible light has a wavelength of about $5 \times 10^{-5} = 0.00005$ meters, an optical microscope *cannot* resolve an object smaller than this distance scale. If we try to examine the parts of the nucleus of the cell of an organism with a microscope in the biological lab, they will appear fuzzy to us, because they are becoming as small as this distance scale. Still smaller objects are not resolvable at all. We can use the most expensive optical microscope money can buy, and still this fuzziness does not go away. This is due to the wave nature of light and the wavelength being larger than the thing we wish to view.

De Broglie taught us, however, that electrons have wavelike properties, and it is fairly easy to make an extremely short-wavelength electron, much smaller than the nucleus of the biological cell. We need only *accelerate* the electron to a large momentum, about the same as produced in a television picture tube. Thus electron microscopes can resolve much smaller features with far greater clarity than optical microscopes. To study still smaller objects at still shorter distances requires higher-momentum probes, hence higher energies. Thus to study the structure of matter within the nucleus of an atom requires a large and powerful particle accelerator—to make probes of the tiniest quantum wavelengths. Particle accelerators and detectors are just enormous microscopes.

THE WAVE FUNCTION

So, if particles can behave like waves, then what is doing the waving?

Suppose we have only one electron in a very large region of space. This is an *approximation* in which one particle is treated as though it is in isolation from everything else in the universe. In fact, this is a pretty good approximation for all freely moving particles, whether they be electrons, light particles (photons), neutrons, protons, or atoms (viewed as particles) when they are straying freely through space (and even, to a lesser degree, within a material, like a metal or a gas).

How do we describe such a lonely particle? Newton in classical physics and Einstein in special relativity would simply say that at time t, the particle is located at a position in space, x. Then the "equations of motion" determine the new position of the particle, x', at some later time t'. This description emphasizes the particle aspect of the object but misses altogether the wave aspect. This description has to be abandoned in quantum mechanics.

Physicists, however, were used to describing (classical) waves in continuous material media, such as sound waves in air (containing many, many particles), long before quantum theory was invented. Consider, for example, an ocean water wave. We describe this with a mathematical quantity representing the amplitude of the water wave, $\psi(x, t)$, (the character ψ is called *psi* and is pronounced like "sigh"). Mathematically, $\psi(x, t)$ is a "function"; that is, it specifies the height of the water wave, relative to sea level, at any point in space x and at any time t. A traveling wave form arises naturally—it is actually the solution to the equations that describe the motion of water when it is disturbed. So, too, are breaking waves, or tsunamis, or any of the many forms and shapes of water waves, all described by a differential equation that determines the "wave function" of water $\psi(x, t)$, the height or amplitude of the water at any point x at time t. We would like to steal the concept of a wave like $\psi(x, t)$ and use it in quantum mechanics. However, when we commit this theft, we become, at first, confused about what we are actually doing.

A mathematically gifted physicist, Erwin Schrödinger, became fascinated with de Broglie's thesis and gave a seminar on the topic in 1924 at his home institution, the University of Zurich. A member of the audience suggested that, if the electron acts like a wave, then there must be a *wave equation* that describes it, just like wave equations that describe water waves.

Schrödinger very quickly had an insight and noticed that the daunting mathematical formalism of Heisenberg could indeed be written in a way that made it look very similar to the familiar equations of physics that describe wave disturbances. Therefore one could say that, at least formally, the correct description of a quantum particle involved a new mathematical function, $\psi(x, t)$, which Schrödinger dubbed the "wave function." Using the machinery of the quantum theory as interpreted by Schrödinger—that is, by solving "Schrödinger's equation"—one can compute the wave function for a particle.[7] At this stage, however, no one yet knew what the wave function of quantum theory represented.

In quantum mechanics, therefore, we can no longer say that at time t the particle is located at a position $\vec{x}$. Rather, we say that the quantum state of motion of the particle is the wave function $\psi(x, t)$, which is the *quantum amplitude* ψ at time t at a position $\vec{x}$. The precise position of the particle is no longer known. Only if the wave amplitude is large at some particular position, $\vec{x}$, and near zero everywhere else, can we say that the particle is localized near that position. In general, the wave function may be spread out in space, like the traveling wave in figure 22, and then we never know, even in principle, exactly where the particle is. Bear in mind, at this stage of the development of this idea, physicists, including Schrödinger, were still very unclear about what the wave function really was.

Here, however, comes a twist in the road that is a stunning hallmark of quantum mechanics. Schrödinger found that the wave function describing a given particle is, like any wave, a continuous function of space and time, but it must take on numerical values that are *not* ordinary real numbers. This is very much *unlike* a water wave, or an electromagnetic wave, which is always a *real number* at each point of space and time. In a water wave, for example, we can say that the height of the waves from trough to crest is ten feet, so the amplitude is five feet, and a small craft advisory will be issued. Or we say that the amplitude at the beach of the approaching tsunami is fifty feet, a huge wave. These are real numbers that can be measured with a variety of instruments, and we all understand what they mean.

The quantum wave function, however, has values for its amplitude that are things called *complex numbers*.[8] For a quantum wave, we would say that at a particular point in space, the quantum wave has an amplitude of $3 + 5i$, where $i = \sqrt{-1}$; that is, i is the number that, when multiplied by itself, gives -1. Numbers that are real plus (real times i) are complex numbers. These would have really distressed Pythagoras. In fact, Schrödinger's wave equation itself always involves $i = \sqrt{-1}$ in a fundamental way, and this is what forces the wave function to be a complex number. This mathematical twist in the road to the quantum theory is inescapable.[9]

This strongly hints that we can *never* directly measure the wave function of a quantum-mechanical particle, since we can only measure, in experiments, things that are always real numbers. The question of interpretation of the wave function thus loomed larger than ever. An accomplished German physicist, Max Born, supplied the answer. Born, who worked in the 1920s with Wolfgang Pauli and Werner Heisenberg at the University of Göttingen, at the same time as Emmy Noether's residence there, provided a physical interpretation of the wave function that has both empowered and haunted quantum mechanics ever since. Born, strongly influenced by Heisenberg's uncertainty principle, proposed that the (absolute) square of the wave function, which is always a real and positive number, is the probability of *finding the particle* at any given point in space at any particular time:

$$|\psi(\vec{x},\ t)|^2 = \text{probability of finding particle at position } \vec{x} \text{ at the time } t.$$

Born's interpretation of Schrödinger's wave function thus locks together, inextricably, the notion of a particle to the notion of a wave. It is also terrifying, or humiliating, depending upon one's perspective— physics must now deal with probability as a *fundamental component of a physical theory*. We can no longer make exact statements about the familiar positions and motions of things. We must be content, according to the very laws of physics, with more limited information about the outcome of a physics experiment. Unlike the language of Newton or Einstein, we cannot talk about the exact position of particle $\vec{x}$ at time t. Rather, all of our available information is now encoded into $\psi(x,\ t)$: the value of the quantum wave function at position $\vec{x}$ at time t, and only its absolute square is measurable. In fact, it was Max Born who coined the term *quantum mechanics*—and he was also the grandfather of pop singer Olivia Newton-John.[10]

Quantum mechanics is inherently a probabilistic theory. The concept that physics, at the atomic level, can predict only probabilities was an enormous philosophical departure from classical physics, so much so that it took years (and tears) before a new intuition was accepted by physicists.

Here is an example. Suppose you are walking on a sunny day along a busy street. You happen to pass a Danish bakery, and you see through the window an assortment of delectably edible Danish pastries. You also happen to see a dim but recognizable image, a reflection of yourself in the store window. What is happening? Sunlight, a stream of photons, strikes your face, and some fraction is reflected and heads toward the window. At the window, many of the photons continue through, illuminating the raspberry cheese swirls inside, but a small fraction of the photons is reflected back toward your eyes, and you see the image of yourself ("Hmm, that's a size 36—maybe I shouldn't stop for a pastry, after all"). This is all plausible classical physics until we ask about a single photon—what decides whether a particular photon will be reflected or transmitted through the glass?

The answer, after solving Schrödinger's wave equation, is shocking: *part of the wave function of a single photon gets through the glass, and part is reflected.* We can therefore say only that the photon has a certain probability of getting through; let's suppose 98 percent for the square of the part of the wave function that gets through and 2 percent for the square of the part of the wave function that was reflected. The photon itself did not break into two pieces—one that went through and one that did not—but the wave function did! The photon definitely gets through the glass or definitely doesn't, but we can only compute the probability of a given outcome, not a definite outcome. The quantum mechanical answer is that, even if we know everything there is to know about glass and photons, and even Danish pastries, we'll never be able to do better than to calculate the probability of the photon being reflected or transmitted through the glass.

Because of this probabilistic nature of physical reality, Einstein never accepted the quantum theory. "In any case I am convinced that He [God] doesn't throw dice," Einstein proclaimed. However, in the case of the photon hitting the bakery window, we see that the decision to reflect or transmit is indeed a throw of loaded dice. In fact, the development of quantum theory was in full swing in the mid-1920s, whereas Einstein's *era mirabilis*, the time of his earth-shaking insights, had essentially ended. All physicists today, however (except for a fringe element), accept the overwhelming validity of the quantum theory.

A BOUND STATE

A particle in classical physics can be trapped by a force into a *bound state.* We have seen this for the Keplerian orbits of a planet orbiting the Sun, where the planets are attracted to the Sun by gravity, or we say they are moving within the gravitational potential around the Sun. We know that something similar happens for atoms, and we get Bohr's discrete states of motion. How does this happen, in terms of the wavelike behavior of the particle?

This is most easily understood by considering a very simple example, an electron bound to a long molecule. It turns out that the form of the wave function of an electron trapped on a long molecule is exactly the same as the shape of the motion of a plucked guitar string. In fact, we can easily work out the energy levels of the trapped electron by thinking about the guitar string vibrations.

We suppose that a quantum particle, let us say an electron, falls into a long, narrow ditch that we call a *one-dimensional potential well.* That is, we suppose that the allowed positions of the electron are constrained, by the various forces of electromagnetism and the arrangement of the atoms in the long molecule, so that it can move only within this restricted domain.

Think for a moment of a ditch of a finite length, L, into which a (classical) tennis ball has fallen. The tennis ball will bounce and roll from one end of the ditch to the other, bumping into the wall at the end of the ditch, then changing direction and rolling back until it hits the other wall. If these collisions were perfectly *elastic*, conserving the kinetic energy of the ball, the ball would roll around inside the ditch forever, bouncing off one wall, then reversing and bouncing off the other. At zero energy, the tennis ball would simply come to rest and sit somewhere in the ditch. Now, however, the tennis ball is being conceptually replaced by an electron, trapped in a tiny but deep ditch, and the quantum effects are now relevant to us.

Now, go get that dusty old guitar out of your closet, provided that it has at least one single remaining guitar string. The guitar string is pinned down at two places, one by the bridge of the guitar and the other by the nut, at the end of the neck of the guitar. When we pluck a string, it vibrates, producing a musical note. The vibrations of the guitar string are *trapped* or *standing waves*. Indeed, if the length of the string were infinite, we could pluck the string and send a traveling wave down the string, off to infinity, representing a free traveling particle in empty space in quantum mechanics. But our guitar string has a fixed length, spanning the nut to the bridge, denoted by L—typically about one meter for the average guitar.

Let's lightly pluck the guitar string at its midpoint, preferably with our thumb, not with a pointed guitar pick. This excites the *lowest mode of vibration* of the string. This corresponds to the lowest quantum energy state of motion of the electron trapped in the ditch. This mode of vibration, we can see from figure 23, has a wavelength that is λ (lambda, pronounced "LAM-duh") $= 2L$, which means the length L is just *half* of an entire wavelength (i.e., there is only one crest or one trough at the peak of the oscillation, whereas a full wavelength would have both a crest and a trough). This is the *lowest mode* or *lowest energy level* or the *ground state* of the system, corresponding to the lowest note of the plucked guitar string. The form of the wave is shown in figure 23.

Now we consider the *second mode* of oscillation of the guitar string. This mode has a wavelength that is now $\lambda = L$. That is, we can have both a crest and trough within the full distance $\lambda = L$, as can be seen from figure 23. You can actually excite the second mode on a real guitar string, with a little patience, by lightly holding your finger to the midpoint of the string while plucking it halfway between your finger and the bridge, then

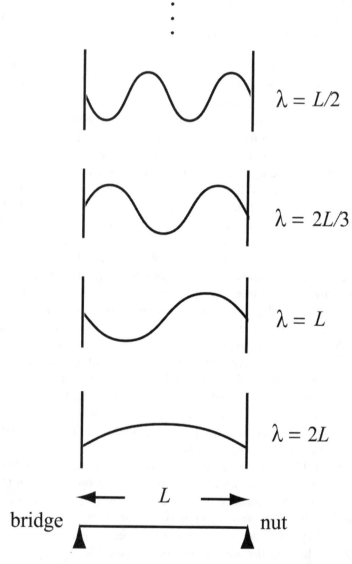

Figure 23. Guitar string representing an electron trapped in a potential well, such as a long organic molecule like beta-carotene. The shape of the guitar string vibrations for each tone possible on the guitar is the same as the shape of the electron's wave function. The electron has increasing energy as the wavelength becomes shorter. The electron could undergo transitions, or hops, between different states of motion, emitting light of a definite energy: the difference of the two energy levels.

quickly removing your finger. The finger assures that the center of the string doesn't vibrate, which we see is a feature of the second mode of oscillation (such stationary points are called *nodes* of the wave function). This produces a pleasant and somewhat harplike, angelic tone, one octave above the lowest mode. Because this has a shorter wavelength, the second mode of the quantum particle has a higher momentum and therefore a higher energy than the lowest mode. If we shine a photon on our electron with just the right amount of energy, we can accelerate the electron and cause it to hop into the second mode, or the *first excited quantum state* of the system. Likewise, the electron can radiate a photon and jump down from this state into the ground state.

The next sequentially higher energy level is the third mode of vibration of the guitar string, which has 1½ full waves, which means that $\lambda = 2L/3$. This can be excited on the guitar by holding one's finger at one-third the length of the string below the nut and plucking at the midpoint of the bridge and finger, then quickly removing the finger—one should then hear a very faint, angelic fifth note (if the string is tuned to C, this note is G in the second octave above C). This corresponds therefore to a still shorter wavelength and a correspondingly large momentum; thus the energy is still larger.[11] Again, a photon hitting our electron with just the right amount of energy can accelerate the electron into this state from the other excited states. Or the electron can radiate a photon and jump down into the lower-energy states from here.

With the application of more energy, we can get the electron to hop to the fourth, fifth, sixth, and higher energy levels, which each correspond to the higher modes of vibration of the guitar string. Eventually, the electron can get sufficient energy to escape the potential, and it then becomes a free particle (its wave function travels away from the scene). We say that the system has become *ionized.*

There are real physical systems that behave exactly like this example of an electron trapped in a one-dimensional ditch. In long organic molecules such as beta-carotene, the molecule that produces the orange color of carrots, the electrons in the outer orbits of some carbon atoms become loose and move over the full length of the molecule, like the electron trapped in a long ditch. The molecule is many atomic diameters long but only one atomic diameter wide. This is shaped very much like our one-dimensional potential—a kind of deep ditch in which the electron moves. The photons that are given off by this molecule when the electrons hop from one quantum state to another have discrete energies corresponding to the difference between two of our energy levels. For long molecules, with a large value for L, these photons correspond to red and infrared light. Measuring the spectrum of emitted photons from similar organic molecules in the lab allows us to determine the length scale L and even to deduce the structure of the molecule.

In general, bound particles, such as electrons in an atom, can hop only between discrete, quantized states of motion. Atoms can thus radiate or absorb photons of definite discrete energies. These discrete spectral lines can be observed with a simple home-built *spectrometer*.[12] We also see that, even in the ground state, the electron, unlike a classical tennis ball, is not at rest. It has a finite wavelength, therefore a finite momentum and therefore also kinetic energy. This ground state motion is called *zero-point motion*, and it occurs in all quantum systems. The electron in a hydrogen atom in the state of lowest energy is moving and not at rest. When we speak of the temperature *absolute zero*, we actually mean a temperature at which everything is in its ground state, not a state of zero motion, since quantum mechanics forces things, even in the ground state, to remain in perpetual motion. Perhaps this vindicates the countless efforts of the many who have attempted to build a perpetual-motion machine—nature *is* a perpetual-motion machine, owing to quantum mechanics. Nonetheless, the conservation of energy and Noether's theorem still hold—the Acme Power Company will not work any better in the quantum regime than it did in the Newtonian world.

The motion of *any* particle that is localized, or trapped within a potential, will behave like the trapped waves on a guitar string and have corresponding energy levels that are quantized and take on only particular, discrete allowed values. This happens to electrons bound in atoms and protons and neutrons bound in the atomic nucleus, as well as to quarks that are bound inside of protons and neutrons. In the case of quarks bound within particles, the energy levels that represent the excited states of motion of the quarks actually appear to us as new particles! And, finally, *string theory* is a relativistically glamorized version of a guitar string. The hope is to explain the quarks themselves (and the other truly fundamental particles in nature) as the quantum vibrations of a string. Such wonderful music can be heard from that old guitar, if one simply practices.

SPIN AND ORBITAL ANGULAR MOMENTUM IN QUANTUM THEORY

Angular momentum is the physical measure of the rotational motion of a system or object. It is the conserved quantity in physics, which comes about, according to Noether's theorem, from rotational symmetry. Angular momentum, which was a continuously varying quantity in Newtonian physics, also changes its character drastically in quantum mechanics. It, too, becomes "digital," or quantized.

Think of a large classical spinning gyroscope. As it spins, it has angular momentum. In classical physics it can apparently have any continuous value of its spin angular momentum we wish. But, as the gyroscope becomes smaller and smaller, we eventually find that the angular momentum is not an arbitrary number, like the average number of children per household, but rather takes on discrete values, like the real number of children per household. Angular momentum is *always quantized* in quantum mechanics. All observed angular momenta, it turns out, will be discrete multiples of Planck's constant divided by 2π, or "h-bar," $\hbar = h/2\pi$. All the particle spin and orbital states of motion we find in nature have angular momenta that can have only the exact values

$$0, \frac{\hbar}{2}, \hbar, \frac{3\hbar}{2}, 2\hbar, \frac{5\hbar}{2}, 3\hbar,$$

and so forth. Angular momentum is always either an *integer* or a *half-integer* multiple of $\hbar$ in nature.

Now, what does this tell us about the angular momentum of the spinning Earth? Essentially nothing. Earth's angular momentum is so large compared to Planck's constant that we can never answer this question to any precision that makes it meaningful. It is also complicated because Earth's total spin angular momentum is not a precise value of a quantum spin. Earth is an enormous system composed of many atoms constantly interacting with its surroundings. The angular momentum of Earth is not precisely determined, viewed on the quantum level. This is not at all like the eternal and stable angular momentum of a single tiny electron. Only at the level of exceedingly tiny systems, atoms, or the elementary particles themselves do we observe the quantization of angular momentum.

In a sense this quantization of angular momentum happens because rotational motion is bounded—much like the electron trapped in a long potential well. We can only rotate a system through 360 degrees (or 2π radians), and then we get the system back to where we started. This is just the rotational symmetry of space. A particle must "live" within a bounded angular space where the angle runs from 0 to 360 degrees (or 2π radians). Therefore, just as the momentum of an electron in a bounded potential well becomes quantized (as well as the energy, which is directly related to the momentum), the analogous quantity, angular momentum, is quantized because of the bounded nature of rotations.

Angular momentum is also an intrinsic property of an elementary particle or an atom. All elementary particles are miniature gyroscopes and have spin angular momentum. We can never slow down an electron's rotation and make it stop spinning. An electron always has a definite value of its spin angular momentum, and that turns out to be, in magnitude, exactly $\hbar/2$. We can flip an electron and then find that its angular momentum is pointing in the opposite direction, or $-\hbar/2$. These are the only two observable values of the electron's spin. We say that the electron is a spin-1/2 particle, because its angular momentum is the particular quantity $\hbar/2$.

Particles that have *half-integer multiples* of $\hbar$ for their angular momentum, that is,

$$\frac{\hbar}{2}, \frac{3\hbar}{2}, \frac{5\hbar}{2},$$

and so on, are called *fermions*, after eminent physicist Enrico Fermi, who helped pioneer these concepts. The fermions that concern us here are the electron, the proton, and the neutron (and a few others later on, like the quarks that make up the proton and neutron, etc.), and each has an angular momentum of $\hbar/2$. We refer to all of these as *spin-1/2 fermions*.

Particles, on the other hand, that have angular momenta that are *integer multiples* of $\hbar$, such as 0, $\hbar$, $2\hbar$, $3\hbar$, and so on, are called *bosons*, after famous Indian physicist Satyendra Nath Bose, a friend of Einstein who also developed some of these ideas. There is a profound difference between fermions and bosons that we'll encounter momentarily. Typically, the only particles that are bosons and that will concern us presently are particles like the photon (called "gauge bosons"), which has "spin-1," or one unit of $\hbar$ angular momentum; the particle or gravity, the graviton, which has yet to be detected in the lab and which has "spin-2," or $2\hbar$ units of angular momentum; and particles called mesons, that have "spin-0," or 0 units of angular momentum. All orbits in quantum theory have integer units of $\hbar$ for angular momentum, hence 0, $\hbar$, $2\hbar$, $3\hbar$, and so on.

THE SYMMETRY OF IDENTICAL PARTICLES

There exist large physical systems in the universe, in the laboratory, and even in your home that are governed in very visible ways by quantum mechanics. In these particular systems the subtle quantum effects are not hidden and are not "averaged to zero."

One of the most amusing of these macroscopically bizarre phenomena is something called a *superfluid*. An example of a superfluid is ultracold liquid helium (in fact, it must be the isotope ^{4}He, and it must be cooled to within a degree or so of absolute zero). If a superfluid is sitting in a glass on a table, it looks like any other liquid, like water, for example. But unlike water, if you slosh a glass of liquid helium, the entire body of liquid will crawl up the side of the beaker, down the other side, and disappear as it evaporates on the floor. You can make a fountain with a superfluid that runs by itself, without a pump. The fluid will climb up a tube and fall back into the beaker and recycle forever. This doesn't violate Noether's theorem or energy conservation, because there is no net energy loss—energy is conserved. We won't think of resurrecting the Acme Power Company, because we can't create any excess energy this way. It is a "super"-fluid in the sense that there is *absolutely no resistance to flow* as a liquid. It is as if the entire body of liquid had become one collective object in one common frictionless state of motion, with all of the atoms of the liquid moving together in concert in exactly the same way, like a grand flock of geese! In fact, that is exactly what is happening: all atoms are moving together in a single state of motion—a spooky quantum mechanical effect generally known as a *coherent state*.

No one has yet found a marketable use for superfluids, but many related quantum systems that use coherent states are now in everyday use and inhabit people's homes. For example, a laser produces a coherent state of light. This is an intense beam of light in which all of the photons, the particles of light, are moving exactly together in lockstep, like a kind of photonic superfluid. The photons are particles, and they are waves—they are both and neither—quantum-mechanical particle-waves that do what quantum mechanics orders them to do. Lasers are the key to devices such as DVD or CD players, in which a large amount of data is stored in a dense optical medium and is "read" by the laser beam. Lasers play an increasingly central role in telecommunications, by transmitting optical signals over glass fibers. This is "quantum optics" coming of age, becoming a substantial component of our gross domestic product and standard of living.

Yet another example of a remarkable large-scale quantum effect is an electrical *superconductor*. This is typically a lousy metallic conductor of electricity at room temperature, such as an alloy of lead (atomic symbol Pb) or nickel (Ni). However, when superconductors are cooled to within a few degrees of absolute zero, the lowest achievable temperature, in which every quantum particle is in its state of lowest energy, an electrical current can flow in a quantum-coherent state. Like the superfluid, there is exactly no resistance to the current flow. In fact, magnets employing superconducting wire are in widespread use in medical imaging devices. Such superconducting magnets were originally invented for large-particle accelerators, in particular the Fermilab Tevatron. In recent years physicists have discovered "high-temperature" superconductors, which can operate at many tens of degrees above absolute zero. These are likely to have a major impact upon our everyday lives in the future, when all of those unsightly high-tension power lines will someday be torn down and replaced by small, underground, high-temperature superconducting cables. Once placed in operation, someday, there will be no loss in transmission of electrical power from the power plant to the consumer!

These bizarre macroscopic effects are a consequence of a symmetry that is of paramount importance in shaping the physical world—the *symmetry of identical particles in quantum mechanics*. This is the bizarre way that quantum mechanics treats particles that are so fundamental that they have no freckles or warts or other identifying body markings, and they absolutely cannot be distinguished from each other. These particular macroscopic effects pertain to the class of particles we identified as *bosons* (particles with an integer spin). Essentially, you can place as many identical bosons into the same quantum state of motion as you wish—and they like it. Photons and ^{4}He atoms are bosons.

We can understand the origin of these effects as a symmetry in the language of Schrödinger's wave function. Let us consider a physical system containing two particles. For example, this could be a helium atom, which has two orbiting electrons, or it could be our potential ditch, containing any two identical particles. In general, we describe the two-particle system by a quantum-mechanical wave function that depends now upon the two different positions of the two identical particles as $\psi(\vec{x}_1, \vec{x}_2, t)$. Again, according to Max Born (and to the torment of Einstein), the (absolute) square of the wave function is the probability of finding our particles at the points in space, and $\vec{x}_1$ and $\vec{x}_2$ at the time t is $|\psi(\vec{x}_1, \vec{x}_2, t)|^2$.

Now consider the act of *exchanging* one particle with the other particle. In other words, we rearrange our system with the swapping of the two positions $\vec{x}_1 \leftrightarrow \vec{x}_2$. Hence the new "swapped" system is described by the wave function $\psi(\vec{x}_2, \vec{x}_1, t)$, where we have simply interchanged the two particles' positions. But is this really a new system or just the original system we started with? That is, is this the wave function describing a new, swapped system, or is it the same original system?

In everyday life, the category of things that we encounter called "dogs" is very large, and no two dogs are identical, even if they both happen to be in the same subcategory (breed), let's say poodles. If we put a poodle into doghouse number 1 and a terrier into doghouse number 2, we'll have a different system than if we put the terrier in doghouse 1 and the poodle in doghouse 2. However, all electrons are precisely identical to each other. Electrons carry only a very limited amount of information. Any given electron fresh from the electron factory is *exactly* identical to any other electron. The same is true of the other elementary particles. Therefore any physical system must be symmetrical, or invariant, under the swapping of one such particle with another. Swapping identical particles in the wave function is a *fundamental symmetry of nature*. In a sense, nature is very simple-minded in the way it treats electrons, in that it doesn't know the difference between any two (or more) electrons in the whole universe.

This "exchange symmetry" of the wave function must leave the laws of physics *invariant* because the particles are identical. At the quantum level this implies that our swapped wave function must give the same observable probability as the original one: $|\psi(\vec{x}_1, \vec{x}_2, t)|^2 = |\psi(\vec{x}_2, \vec{x}_1, t)|^2$. This condition, however, implies two possible mathematical solutions for the effect of the exchange on the wave function:

$$\text{either } \psi(\vec{x}_1, \vec{x}_2, t) = \psi(\vec{x}_2, \vec{x}_1, t) \text{ or } \psi(\vec{x}_1, \vec{x}_2, t) = -\psi(\vec{x}_2, \vec{x}_1, t).$$

That is, we notice that the exchanged wave function can in principle either be *symmetrical*, that is, +1 times the original, or else it can be *antisymmetrical*, or −1 times the original. Either case is allowed, in principle, because we can measure only the probabilities (the squares of wave functions).

So, which is it, +1 or −1? In fact, quantum mechanics mathematically allows both possibilities, so nature finds a way to offer both possibilities! And the result is astonishing.

It turns out that when we are talking about bosons the rule is that, upon swapping two particles in the wave function, we would get the + sign:

Exchange symmetry of identical bosons: $\psi(\vec{x}_1, \vec{x}_2, t) = \psi(\vec{x}_2, \vec{x}_1, t)$.

With this result, we can immediately anticipate an important physical effect—two identical bosons can readily be located at the same point in space, that is, $\vec{x}_1 = \vec{x}_2$! Therefore $\psi(\vec{x}, \vec{x}, t)$ can be nonzero! In fact, by considering lots of bosons localized in the same region of space, described by one big wave function, we can actually prove that the most probable place for all the bosons in a system is *piled on top of one another!* That is, from just a probabilistic point of view, it is possible to coax a lot of identical bosons to share the same little region in space, in fact almost an exact pinpoint in space. Or we can likewise coax many identical bosons into a quantum state, each boson having the exact same value of momentum. Thus we say that bosons like to "condense" into compact, or coherent, states. This is called *Bose-Einstein condensation.*

As we mentioned above, there are many variations on Bose-Einstein condensation and all kinds of phenomena that have in common many bosons in one quantum state of motion. Lasers produce coherent states of many, many photons all piled into the same identical state of momentum, moving together in exactly the same way at the same time. Superconductors involve pairs of electrons bound by crystal vibrations (quantum sound) into spin-0 bosonic particles. In a superconductor the electric current involves a coherent motion of many, many of these bound pairs of electrons sharing exactly the same state of momentum. Superfluids are quantum states of extremely low-temperature bosons (as in liquid ^{4}He, mentioned earlier) in which the entire liquid condenses into a common state of motion and thus becomes completely immune to friction. Superfluids have to be ^{4}He, because the isotope ^{3}He is not a boson (it is a fermion; see below). Bose-Einstein condensates can occur in which many bosonic atoms condense down into ultracompact droplets of very large density, with the particles piling on top of one another in space. A Bose-Einstein condensate is reminiscent of a football tackle on a wintry Sunday afternoon in Green Bay.

If, on the other hand, we exchange a pair of identical electrons (fermions) in a quantum state, the rule is that we get the minus sign in front of the wave function. This holds for any particle with fractional spin, such as the electron, with spin-1/2:

Exchange symmetry of identical fermions: $\psi(\vec{x}_1, \vec{x}_2, t) = -\psi(\vec{x}_2, \vec{x}_1, t)$.

We can therefore see a simple yet profound fact about identical fermions: no two identical fermions (with spins, quark colors, etc., all "aligned") can occupy the same point in space: $\psi(\vec{x}, \vec{x}, t) = 0$. This follows from the fact that if we now swap the position $\vec{x}$ with itself, we must get $\psi(\vec{x}, \vec{x}, t) = -\psi(\vec{x}, \vec{x}, t)$, and therefore $\psi(\vec{x}, \vec{x}, t) = 0$, because only 0 equals minus itself!

More generally, no two identical fermions can occupy the same quantum state of momentum, either. This is known as the *Pauli exclusion principle*, named after brilliant Austrian-Swiss theorist Wolfgang Pauli. Pauli actually proved that his exclusion principle for spin-1/2 comes from the basic rotational and Lorentz symmetries of the laws of physics. It involves the mathematical details of what spin-1/2 particles do when they are rotated. Swapping two identical particles in a quantum state is identical to rotating the system by 180 degrees in certain configurations, and the behavior of the spin-1/2 wave functions then gives the minus sign.

EXCHANGE SYMMETRY, STABILITY OF MATTER AND ALL OF CHEMISTRY

The exclusion property of fermions largely accounts for the stability of matter. For spin-1/2 particles there are two allowed states of spin, which we call "up" and "down" (up and down refer to any arbitrary direction in space). Thus, in an atom of helium, we can get two electrons into the same lowest-energy orbital state of motion, where the lowest-energy orbital is the analogue of our lowest mode of the plucked guitar string (the ground state of the bound electrons in a potential well). To get the two electrons in one orbital requires that one electron has its spin pointing "up" and the other having spin pointing "down." However, we *cannot* then insert a third electron into that same orbital state, because its spin would be the same, either up or down, as one of the two electrons already present. The exchange antisymmetry would force the wave function to be zero. That is, if we try to exchange the two electrons whose spins are the same, the wave function would have to equal minus itself and must therefore be zero! Hence, for the next atom in the periodic table, lithium, the third electron must go into a new state of motion, that is, a new orbital. Thus lithium has a *closed inner orbital*, or "closed shell" (i.e., a helium state inside of it) and a sole outer electron. This outer electron behaves much like the sole electron in hydrogen. Therefore lithium and hydrogen have similar chemical properties.

We thus see the emergence of the periodic table of elements. If electrons were not fermions and did not behave this way, every electron in the atom would rapidly collapse down into the ground state, and all atoms would behave like hydrogen gas. The delicate chemistry of organic (carbon-containing) molecules would be impossible. There would never be Bach cantatas—or anyone around to hear them.

Yet another extreme example of fermionic behavior is that of the neutron star. A neutron star is formed as the core of a giant supernova implodes while the rest of the star is blown to smithereens. The neutron star is made entirely of gravitationally bound neutrons. Neutrons are fermions, with a spin of 1/2, and again the exclusion principle applies. The state of the star is supported against gravitational collapse by the fact that it is impossible to get more than two neutrons (each with spins counteraligned) into the same state of motion. If we try to compress the star, the neutrons begin to increase their energies because they cannot condense into a common lower-energy state. Hence there is a kind of pressure, or resistance to collapse, driven by the fact that fermions are not allowed into the same quantum state.

In fact, a neutron star often traps the magnetic field of the parent Titanic star that produced it in a supernova explosion. This intense magnetic field then rotates with the neutron star at a high rotational frequency, perhaps hundreds of times per second. This field sweeps through matter orbiting the tiny star, which becomes electromagnetically excited, which in turn produces the phenomenon of rapid flashes of light seen emanating from the star. This is known as a *pulsar*.[13]

Remarkably, if the mass of the neutron star exceeds about 1.4 solar masses, then gravity actually beats out the fermionic exclusion. When gravity wins, the neutron star collapses into—you guessed it—a black hole. Likewise, when stars like our own Sun die, much more peacefully than a Titanic supernova, they cool off, becoming redder and redder, like the dying coals of a campfire. They shrink, initially unable to support themselves with their radiation pressure against gravity. However, eventually the electrons in the star support it against the gravitational collapse. Trying to compress the star forces electrons into higher and higher energy levels, and the star is not massive enough to overcome this "exclusion pressure." Stars with masses less than 1.4 times our Sun's mass will end as lifeless, cold dead worlds called *dwarfs*, as quantum mechanics of matter wins against gravity. Heavier than 1.4 solar masses, and gravity wins—the star becomes squeezed and collapses into a black hole, a dangerous threat to interstellar navigation. This crucial number, 1.4 times the solar mass, is known as the *Chandrasekar limit* and marks the crossover from the exclusion principle to gravity winning the war that determines the ultimate fate of a dying star.

All of these bizarre macroscopic phenomena come from the *exchange symmetry* of the quantum wave functions of elementary particles. We don't observe this exchange symmetry in the case of poodles, or people, or any other everyday macroscopic objects. This is "simply" a consequence of their complexity. Complexity requires that the individual particles have to be far apart from one another, so that many different physical states are possible, and the particles never come at all close to being in the same quantum states at the same time. One poodle differs from another because of this complex arrangement of its quantum components. Thus the effects of identity are not obvious in complex extended systems, far removed from the quantum ground state.[14]

QUANTUM THEORY MEETS SPECIAL RELATIVITY: ANTIMATTER

What happens when quantum theory meets special relativity? Something rather incredible.

We recall from relativity that the energy, momentum, and mass of a particle are related by the formula $E^2 - p^2c^2 = m^2c^4$. This is fundamentally a consequence of the space-time symmetry of special relativity and Noether's theorem. To compute the energy of a particle, we first write, equivalently, $E^2 = m^2c^4 + p^2c^2$, and then, to get the energy, we have to take the square root of this mathematical expression for E. However, every number has *two* square roots. For example, the number 1 has $\sqrt{1} = 1$ and $\sqrt{1} = -1$. That is, we know that $1 \times 1 = 1$ and also that $-1 \times -1 = 1$. The "other" square root of a positive number is negative. How do we know that the energy we derived from Einstein's formulae should be positive? How does nature know? What is the fate of the negative energy solution?

Common sense would tell us that energy, especially the rest energy of massive particles, mc^2, must always be positive. Hence, physicists in the early days of special relativity simply refused to talk about the possibility of the negative square root, saying that it must be "spurious" and that it doesn't describe any physical particles.

But could such *negative-energy* particles exist, particles in which we take the negative square root, $E = -\sqrt{m^2c^4 + p^2c^2}$? With zero momentum, these particles would have negative rest energy of $-mc^2$. Their energy would actually decrease if their momentum were increased. They would continually lose energy by collisions with other particles, and by radiating photons, and they would actually *speed up*! In fact, their energy would become more and more negative, eventually becoming infinitely negative. Such particles would continuously accelerate, and fall down into an abyss of negative infinite energy. The universe would be full of these infinite-negative-energy oddball particles.

This problem is deeply buried in the fabric of special relativity, and it simply cannot be ignored. It becomes more severe when we try to make a quantum theory of the electron. It turns out that we can never avoid the negative sign of the square root. Quantum theory forces electrons to have both positive and negative energy values for any given value of the momentum. We would say that the negative-energy electron is just another allowed quantum state of the electron. But this would be a disaster as well, since it would mean that ordinary atoms, even simple hydrogen atoms, could not be stable. The positive-energy electron could emit photons, adding up to an energy of $2mc^2$, and become a negative-energy electron and begin its descent into the abyss of infinite negative energy. Evidently, the whole universe could not be stable if negative-energy states truly existed. The problem of negative-energy electron states was a prime headache for early attempts to construct a quantum theory of electrons interacting with light that was consistent with Einstein's special theory of relativity.

One day in 1926 a brilliant theoretical physicist, Paul Dirac, had an idea. As we have seen, the Pauli exclusion principle says that no two electrons can be put into exactly the same quantum state of motion at the same time. That is, once an electron occupies a given state of motion—a quantum state—that state is *filled*. No more electrons can join it.

Dirac's idea was that the *vacuum itself* is completely filled with electrons, occupying all of the negative-energy states. If all of these negative-energy states are filled, then positive-energy electrons, such as those found in atoms, cannot emit photons and therefore cannot drop down into these states, because they are *excluded* from doing so. In effect, the vacuum becomes one gigantic inert atom, with all of the possible momentum states of negative energy already filled. This would seem to put an end to the issue of the negative energy levels once and for all.

However, Dirac realized that the story didn't end there. It would be theoretically possible to "excite" the vacuum. This means that we can engineer a collision in which we pull a negative-energy electron out of the vacuum, much like a fisherman pulling a deep-sea fish into the boat. For example, suppose an intense gamma ray collides with an electron occupying a negative-energy state in the vacuum. We also provide other particles participating in the collision, like a nearby heavy atomic nucleus, to conserve momentum, energy, and angular momentum. The gamma ray knocks the electron out of the negative-energy state and into one of positive energy, and recoiling off the heavy nucleus. This would leave behind a *hole* in the vacuum.

Dirac realized that the hole is the absence of a negative-energy electron. This means that the hole actually has *positive* energy. However, the hole would also be the absence of a negative electrically charged electron, hence also a *positively charged* particle. We call this object a *positron*. The hole, at rest, therefore must have an energy of exactly $E = +mc^2$, where m is exactly the electron mass. Positrons are the antiparticles of electrons, and they must exist if special relativity and the quantum theory are both true.

In fact, positrons were discovered in 1933 by Carl Anderson. They were seen as tracks in a *cloud chamber* with a strong magnetic field, causing the particle motion to curve in a way that revealed the electric charge. A cloud chamber was an early kind of particle detector that contained air supersaturated with a water or alcohol vapor. As a charged particle traveled through the chamber, the vapor condensed into little cloud droplets that marked the path of the particle, which could be photographed. Anderson observed pairs of electrons and positrons as two separate curling tracks in a cloud chamber several years after Dirac's theory had predicted them. The mass of a positron is indeed the same as that of an electron, as the symmetry of special relativity requires.

Antimatter will annihilate matter when the two collide, producing a lot of energy (direct conversion of rest-mass energy) in the form of other particles. The electron simply jumps back down into its hole in the "Dirac Sea," and the energy emerges as mostly photons and other low-mass particles.

It would be nice if we could mine antimatter from somewhere in the universe, because it would provide an excellent energy source. However, for reasons that remain mysterious today, there is no abundant source of antimatter left in the universe. As we saw in chapter 8, theorists understand how this could have happened in principle, through CP violation, but for various reasons the precise mechanisms must lie in new physics that has not yet been discovered. Positrons are naturally produced from the radioactive disintegration of certain nuclei and have found a use in medical imaging (positron emission tomography, or PET scans). It is unclear if the utility of antimatter will expand to warp-drive starship engines, but eventually it will surely find some more practical applications, perhaps having great impact upon the future economy.

One far-fetched scheme to solve the nation's energy problems would be to build a particle accelerator in a very close orbit around the Sun where solar energy is abundant (a million miles above the Sun's surface, there are ten megawatts of power per square meter; unfortunately, finding materials that don't melt would be a challenge). With this facility, using the intense solar energy, we could make and collect 500 kilograms per year of antimatter and ship it back to Earth in magnetic bottles. Annihilating it with matter back on Earth would yield the rest-mass equivalent of 1,000 kilograms of energy—the current annual energy demand of the United States. There may be some technical hurdles to this, but presumably none that money can't solve—it's better living through particle physics.

We don't know what the ultimate practical applications of antimatter will be—but we're sure that one day the government will tax it.

chapter 11

THE HIDDEN SYMMETRY OF LIGHT

Aye, I suppose I could stay up that late.

—James Clerk Maxwell,
on being told upon his arrival at Cambridge University that there
would be a compulsory 6 AM church service

It has been known for several hundred years that electric charge is conserved in any physical process. The idea first took hold in the mid-1700s with people like William Watson and Benjamin Franklin. This conservation law is fundamental to the classical theory of electric and magnetic fields, or electromagnetism. We see an example of electric charge conservation when we consider the decay of the neutron, $n^0 \rightarrow p^+ + e^- + \bar{\nu}^0$. The neutron is electrically neutral, having zero electric charge. When it decays, we are left with a positively charged proton, a negatively charged electron, and a neutral (anti)neutrino. The positive charge of the proton identically equals the opposite of the negative electron charge, and the neutrino has zero electric charge, so the final products of the neutron decay have a zero total electric charge. Electric-charge conservation is an exact conservation law in *all* physical processes—we have never seen a net gain or loss of electric charge in any physical process. Given the existence of this conservation law, by Noether's theorem, we are compelled to ask, What is the underlying continuous symmetry that leads to this conservation law?

Electromagnetism, or "electrodynamics," is the physical description of electric and magnetic fields, as well as electric charges and currents, and it was formulated in a classical (nonquantum) framework over the entirety of the nineteenth century. The pinnacle achievement is usually considered to be the formulation of Maxwell's equations, published in 1865 by James Clerk Maxwell, a succinct and complete set of equations that summarize all known aspects of electrodynamics and allow us to compute the electric and magnetic fields anywhere in space and time, given any choice of electric-charge and electric-current distributions.[1]

Maxwell, born in Scotland and living only to age forty-eight, is a towering figure in the history of science. His importance in the history of physics is comparable to that of Einstein and Newton. He was the first to recognize that light is a propagating wave disturbance of electric and magnetic fields and is a solution to the equations that describe all electric and magnetic phenomena, Maxwell's equations. The laws of special relativity are already contained in Maxwell's theory—Einstein "simply" unearthed them by contemplating the symmetries of the equations under different states of inertial motion.

Maxwell's classical theory of electrodynamics makes no sense without the conservation law of electric charge. The underlying continuous symmetry that leads to this, however, appeared at first to be somewhat mysterious and obscure.

Electric charges are the sources of *electric fields*, much as mass is the source of a gravitational field in Newton's theory of gravity. An electric field is just the electric force exerted on a unit electric charge at any point in space. When electric charges move, they become electric currents and produce *magnetic fields*. Magnetic fields in turn produce forces on moving electrons (electric currents). In fact, a pure electric field in space becomes a combined electric and magnetic field if we simply move through it.

The Maxwell theory does not allow solutions to its equations in which a source or a sink, an electric charge, simply disappears into nothingness. This is such a stringent requirement on physics that not even the ancient Greek mythological hell, Tartarus, or a black hole can cause an electric charge to disappear. If an electric charge falls into a black hole, the black hole itself will have the same value of the electric charge that it swallowed. If we stop the discussion at this level, however, it would be incomplete—what is the underlying continuous symmetry required by Noether's theorem that causes electric charges to be conserved? It must be there somewhere, but where?

HINTS OF A SYMMETRY

If we probe deeper into the mathematical structure of Maxwell's theory, however, we find that there is something even more fundamental than the electric and magnetic fields. This more fundamental thing was given a fancy name—it was called a *gauge field*. The gauge field is related to the electric and magnetic fields in a peculiar way: if we are given the gauge field in any region of space and time, we can always calculate the values of the electric and magnetic fields in that region. However, we cannot reverse this process. That is, given electric and magnetic fields in the same region of space and time, we cannot determine exactly what gauge field produces them. In fact, we can always find an *infinite number* of gauge fields that would produce the same observed electric and magnetic fields.

The gauge field will always be undetermined—there is always an ambiguity in the form of the gauge field if we try to reconstruct it. Moreover, whereas electric and magnetic fields are easily measured in the lab, we cannot determine the gauge field by either theory or experiment. Even a zero value everywhere for the electric and magnetic fields—that is, a vacuum—does not determine the value of the gauge field—infinitely many different gauge fields exist that yield zero values of the electric and magnetic fields. The gauge field is therefore a "hidden field," not amenable to any measurement that can determine its exact form.

The concept of a gauge field was first considered as a tool for conveniently expressing electric and magnetic forces, by various scientists in the early to mid-1800s. Often different people would write down different gauge fields, in different forms, and it was always unclear whether or not they were describing different phenomena. In 1870, Hermann Ludwig Ferdinand von Helmholtz, a famous contributor to the theory of electromagnetism, showed that different forms of gauge fields can lead to the same physical consequences, that is, to the same electric and magnetic fields. Hence, one can continuously transform one gauge field into another, and the physics stays the same. This is essentially the first example of a new symmetry transformation of electrodynamics—a "gauge transformation"—though its implication as a fundamental symmetry of nature was not appreciated at the time.[2]

In fact, if we turn this around and we insist, as a symmetry principle, that the gauge field must always be a hidden field and can never be determined unambiguously, we then discover something remarkable: gauge symmetry implies that electric charge must be conserved! We can continuously *transform* our chosen gauge field into another one, without changing the values of the electric and magnetic fields, and this is the symmetry that, by way of Noether's theorem, leads to the conservation of electric charge. This weird, hidden symmetry is called *gauge invariance.*

Hidden fields, or "hidden variable theories," have always been psychologically disturbing to physicists. Many scientists over the ages have argued on philosophical grounds against them—nature should be strictly describable in terms of the things that we can directly measure or observe. This notion seems to derive from philosophers such as Descartes, who argued against hidden demons manipulating the world in unseen ways. Yet evidently this philosophical issue doesn't bother nature. The full quantum wave function of an electron itself is not directly observable—only its absolute magnitude, the probability of finding the electron located somewhere, can be measured in an experiment. Now the gauge field joins the wave function as an unobservable phenomenon of nature.

But wait—can these two hidden attributes of nature be fused together to make something grander still? Indeed, the new symmetry of gauge invariance becomes much more compelling and, in a sense, easier to understand, when we enter quantum mechanics. It is as if classical electrodynamics begs for the existence of quantum mechanics.

LOCAL GAUGE INVARIANCE

It was in the twentieth century, with the development of quantum mechanics and the effort to include both the electron and electromagnetism in one completely consistent theory, that the symmetry of gauge invariance emerged as the overarching theme. In fact, this has been the dominant theme in all of twentieth-century physics—*all* forces are now known to be governed by gauge symmetries, descriptions of which are called *gauge theories.*

We recall that all particles are described in quantum theory by waves, through their wave functions. The information about the particle's momentum is determined by the wavelength of the wave, and the energy by the frequency, through the formulae $E = hf$—energy equals Planck's constant times the frequency—and $p = h/\lambda$—momentum equals Planck's constant divided by the wavelength. We have seen that, despite the fact that this energy and momentum information is always present in the wave function and can be extracted from it, we can never measure the wave function directly, because the wave function involves complex numbers that don't make sense as physical observables. Max Born argued that only the (absolute squared) magnitude of the wave function, which is the *probability*, can actually be measured.

Let's investigate this hidden quality of the electron's wave function in greater detail. We'll consider an electron trapped in a large room, and the electron's wave function thus fills the entire room. One way to think of the quantum wave function of the electron is to pretend that we have a special instrument, an "Acme wave function detector," that can measure the quantum wave function of the electron in its entirety. The detector has a circular dial, a *gauge*, with an arrow indicator that points to numbers on the dial. The numbers on the dial are like those of a clock, and the arrow is like the long hand of a clock. Our Acme detector also has an indicator light that can glow bright or dim. With this detector, we imagine that we can measure the full wave function of the electron, the thing we called $\psi(x,t)$ in the previous chapter, by walking around space (and time) and looking at the dial and the indicator light.

For the Acme detector, the brightness of our imaginary indicator light is the probability of finding our electron at any point in space or time. This is what Max Born identified as the absolute square of the wave function, $|\psi(x,t)|^2$. The probability of finding an electron somewhere in space is physically observable in experiments—and wherever our indicator light is glowing brightly is where we are most likely to find the electron. So the brightness of the indicator light corresponds to something that is measurable in nature, not hidden from view by nature.

However, our detector also tells us something else, that is, the particular number on the dial that the arrow points to. This is called the *phase* of the wave function, and although our detector can measure it, the phase of a quantum wave function is *not* directly observable by any other means. Nevertheless, observable information about the energy and the momentum of the electron is encoded into the phase.

For example, let us stand at one location in space in the room, perhaps where the indicator light is glowing a medium brightness, so the electron has a finite probability of being there. We look at the arrow indicator. We happen to see that the arrow is spinning around the dial, from 12, to 3, to 6, to 9, and back to 12, completing a full cycle around the dial once every second. This is the *frequency* of the electron's wave function, f = once per second, and from it we can deduce the energy of the electron from Max Planck's formula, $E = hf$. Suppose also that we have many little detectors arrayed along a straight line in space. At one instant in time we look at all the arrows on each of the different dials. The first arrow points to 12 on the dial, the next detector's arrow to 3, the next to 6, the next to 9, and the next to 12 again. This full cycle happens over, let's say, a distance of 10 meters. Therefore we have measured the wavelength of our electron's wave function, λ = 10 meters, and this determines the momentum (actually, the vector component of the momentum along the line we've chosen in space) to be $p = h/\lambda$. Remember, although our imaginary Acme detector can read it, the phase of the wave function is actually hidden from us in the real world.

A bright young student sitting in the front row asks, "This all seems really strange to me. What would happen, for example, if we somehow changed the electron's wave function *without* changing the observable probability at any point in space and time? Let's make the unobservable phase of the electron's wave function completely different but keep the probability the same at every point in space. How could this be a different physical state of an electron if we can't observe anything different about it? Or is it really the same? Could this be a symmetry transformation of an electron that we haven't yet thought of?" She is thinking in terms of transformations that leave things the same—symmetry, the lesson of this book. She gets an A.

So, somehow we imagine that we can make a change in the electron's wave function that keeps the brightness of the indicator light the same at every point in space and time, so the observable probability of finding our electron isn't affected. But we do change the electron's wave function such that the arrow (or phase) now points randomly to any value on the dial as we move through time or space. If we stand still, the arrow now moves around in time, continuously, but there is no regular frequency—the arrow now moves smoothly from 12 to 1, then smoothly counterclockwise back to 9, then stops and moves forward clockwise to 6, then to 8, and so forth. We seem to have changed the electron's quantum state in a significant way, since it apparently now has no definite frequency, hence no definite energy, even though the indicator brightness is just the same as before. As we walk around the room the light still becomes bright where it is most probable to find the electron, and dim where the electron is unlikely to be found just as before we made the change.

Since the effect of this change in the electron's wave function is only affecting the behavior of the arrow of our detector—that is, the "gauge" part of our detector—we call this a *gauge transformation*. But, in making this change, there is apparently nothing invariant here. This is evidently not a symmetry of the original quantum state, but rather it seems to produce a new quantum state with a different observable energy and momentum.[3] This is shown in figure 24. Here we change just the wavelength of the incoming electron, that is, the distance in space over which the detector arrows go around one full cycle on the dial face. This evidently causes the momentum of the incoming electron to be physically changed, so how can this possibly be a symmetry?

But let us now suppose that there is some other quantum particle in the world that interacts with our electron. And let us further suppose that when we change the electron's wavelength or frequency, we simultaneously create a quantum state that also contains this new particle. The new particle has a wave function that is just a field in space and time, in which the electron moves but with which the electron interacts. What is the effect of this additional new field?

Looking closer at our Acme wave function detector, we see that there is a little switch that, when flipped on, detects the effect of the gauge field. We flip the switch on, and we again observe the electron's wave function, with the modified phase. Indeed, the arrow still seems to move about in the random way from 12 to 1, then smoothly counterclockwise back to 9, then stopping and moving forward clockwise to 6, then to 8, and so forth. But looking closer, we now see that the dial also rotates! So, while the arrow changed from 12 o'clock to the 1 o'clock *position*, the dial itself rotated backward two notches, so that the arrow now points to 3 on the dial, exactly as it did before. Both the dial and the arrow are in a different position, yet the meter is reading 3! As we continue to watch the electron, we see that the arrow now moves to the 9 o'clock position, but the dial also moves forward five notches, so the arrow now points to the number 6 on the dial, exactly as it did before we changed the electron wave function. Watching the actual number that the arrow points to on the rotating dial, we see that it goes from 12, to 3, to 6, to 9, and back to 12 again, completing one full cycle in one second! The frequency, f, of the electron, when the gauge field is included, is exactly what it was before, one cycle

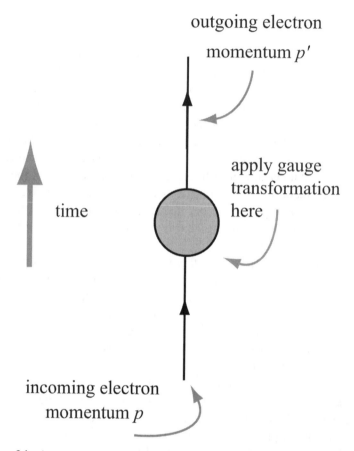

Figure 24. A gauge transformation on an electron's wave function with momentum p changes the wavelength of the wave function, and hence the electron, to a different momentum, p'. Without the gauge field this transformation is *not* a symmetry, because the final electron state is different from the initial one.

per second. Thus the energy of the electron is also exactly what it was before, given by Planck's formula $E = hf$.

This is shown schematically in figure 25. Together with the new gauge particle, we maintain both the original incoming energy and momentum, even though we scramble the unobservable phase of our electron's wave function. Thus the term *gauge* means that the actual determination of the physical momentum of the electron requires the presence of the calibrating gauge field. Only the electron wave function, together with the gauge field, yields a physically meaningful total momentum and energy.

The presence of the new field interacting with the electron is designed to *compensate* the change in our electron wave function, resetting the total momentum, electron plus the effect of the gauge field, back to the original momentum of our original incoming electron.[4]

This is indeed a spooky concept. Acme wave function detectors don't really exist. We emphasize that the gauge field is not observable; the electron wave function is also not observable. We are performing an unobservable transformation of two unobservable objects! Even if these things were directly observable, this says that the very meaning of an electron, by itself, is not absolute. An electron is equivalent, by a gauge symmetry transformation, to a different electron with a different wavelength,

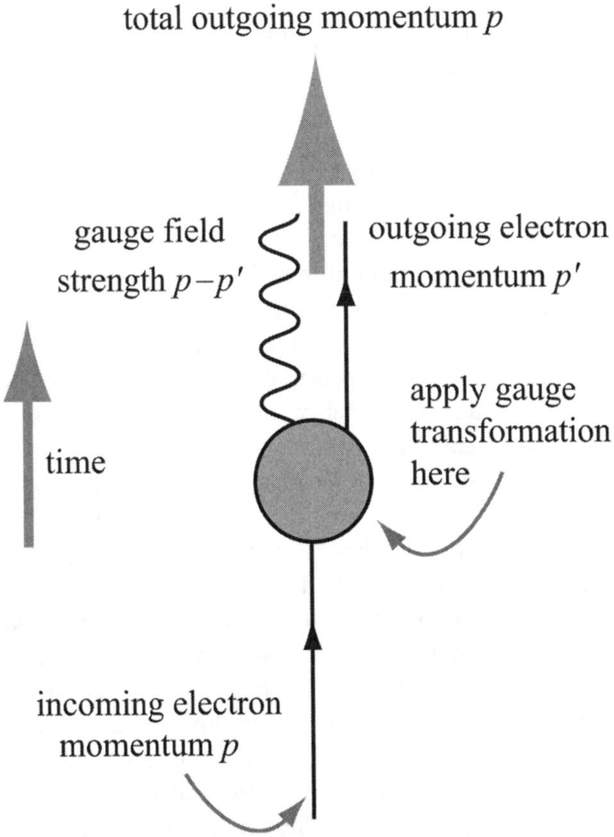

Figure 25. A gauge transformation on electron wave function, together with the gauge field, keeps the total momentum and energy of the system the same. This is now a symmetry. The electron wave function is always a "blend" of the pure mathematical wave function together with the gauge field.

together with the gauge field that resets the total momentum at its original value. The electron and the gauge field are effectively blended together to make one symmetrical entity. The question is, have we really done anything at all, or have we just been playing games? What is this spooky gauge field? Is there any physically observable content to this weird new symmetry?

Yes, there is: when She invented gauge symmetry, God said, "Let there be light."

THE QUANTUM PROCESS OF RADIATION: QED

The gauge theory with its hidden symmetry leads to a profound consequence: if the electron is given a physical "kick," that is, if the electron is *accelerated*, then the gauge field itself is actually emitted as a quantum particle. To visualize this, suppose now we give our initial electron a real physical kick. If the electron initially has a momentum p, and we kick it into a new state with momentum k, then we will produce a physical "gauge particle" with a momentum $p - k$. The gauge field, the spooky thing that previously only haunted the electron, has now become a true physical entity, separated from the electron as it accelerates, and it is radiated as a physical quantum particle out into space. Indeed, the gauge field is now a propagating wave of (measurable) electric and magnetic fields with its own momentum and energy, just as Maxwell had envisioned light, and it can be detected far away as a real photon. It is as though the electron, in having been accelerated, has shaken off the ghostly gauge field that surrounded it. From the point of view of a distant observer, an accelerated electron has radiated a new particle that is physically detectable. This new particle is called the *photon* (see fig. 26).

Light is therefore emitted from accelerated charges. Indeed, this occurs in countless many physical processes, such as the *scattering* of an electron off of an atomic nucleus, or an atom, or another electron. It can be observed readily in the laboratory. At very low energies, it is the way in which electrons emit photons from a campfire. Accelerated electrons radiate the microwaves that heat our coffee in a microwave oven, or transmit the evening news into our living rooms, or cause the Sun to shine.

The emission of a photon

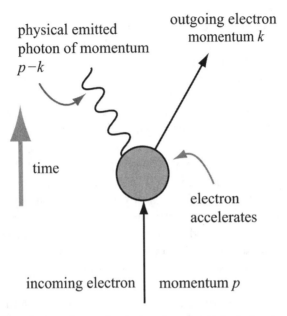

Figure 26. The electron is accelerated and a physical photon is radiated into space, carrying detectable momentum and energy.

FEYNMAN DIAGRAMS

The dynamics of gauge theories such as electromagnetism (or any interactions in quantum mechanics, for that matter) is most readily visualized in terms of *Feynman diagrams*. These diagrams are not just mere pictures of a physical process. They tell us precisely how to compute the quantum outcome, the probability of a given process, provided that the strengths of the interactions are known and are not too large. We can illustrate conceptually how it works. Let us consider a typical process: the scattering of two electrons off of each other by the electromagnetic force. Feynman diagrams show how this arises at the quantum particle level.

The law of electromagnetic force between two charged particles was first proposed in the late eighteenth century by Charles-Augustin de Coulomb and bears a striking resemblance to Newton's law of gravitation. The force between two static charged particles is an inverse-square law force. With two electric charges, q_a and q_b, at rest, located at a separation of R, the *potential energy* due to this force between the charges is $k\,q_a\,q_b/R$, where $k = 9.0 \times 10^9$, and the charges are measured in coulombs. The electric charge of the electron is negative and is determined to be $q_{electron} \equiv -e = -1.6 \times 10^{-19}$ coulombs.

Now, charged particles are generally moving fast, often near the speed of light. The static theory of Coulomb is thus of limited use to describe them. The full classical theory of Maxwell allows us to accommodate the electron motion near the speed of light, but it treats the electron as a pointlike classical particle, and light as a classical wave, and it doesn't contain any quantum physics. We know, however, that the photon and electron are both quantum particles, behaving simultaneously as waves and particles. It thus became essential to have a complete description of the interaction of electrons with photons, perhaps the most important fundamental interaction in all of physics, biology, and chemistry, correctly incorporating all the laws of physics into one theory. The modern, fully relativistic quantum theory of electrons interacting with photons is known as quantum electrodynamics (QED), and it completely and beautifully solved these problems. In fact, this is the most precisely and thoroughly tested theory we have in all of physics—and probably in all of human knowledge.

The problem of formulating QED and making it useful was solved independently by Julian Schwinger, Richard P. Feynman, and Sin-Itiro Tomonaga in the late 1940s, for which they shared the Nobel Prize in 1963. Schwinger's approach was one of brute mathematical force. He developed many powerful and sophisticated techniques that now underlie all of quantum field theory and have even improved our understanding of the classical theory of electromagnetism. Much of the development of the powerful electromagnetic radiation produced by such advanced instruments as synchrotron light sources is based upon Schwinger's work. These are intense sources of gamma rays that can analyze the rapid time dependence of subtle processes in chemical reactions, the structures of metals, the properties of rare nuclei, and even the physics inside a fusion reactor. And Schwinger was first to compute some of the subtle electromagnetic properties of the electron, obtaining the first dramatic result for the quantum corrections to the magnetic field of the spinning electron (called the anomalous magnetic moment).

Feynman, on the other hand, adopted a more intuitive approach to the problem of QED and created a completely novel way of computing quantum interaction effects. This has become the most illuminating technique employed in virtually all subdisciplines of physics today. Here we graphically represent a physical process that also represents the quantum computation by the use of Feynman diagrams. We can often visualize a process through these diagrams even when we cannot compute the result. A graduate student writing from Cornell University, where Feynman developed this technique, commented, "At Cornell, even the janitors use Feynman diagrams."

Consider two electrons that collide and scatter off one another by the force of electromagnetism. This process is represented in figure 27 by a Feynman diagram. The idea is that the process of quantum scattering is determined by something called the T-matrix, from which we can compute the probability of the process happening by taking the absolute square of T, $|T|^2$. The T-matrix is directly related to the total potential energy of the pair of electrons when the electrons reach the point of closest approach, just as we wrote above in the classical theory.

The simplest Feynman diagram actually rediscovers the old classical result of Coulomb when the two electrons are essentially at rest or moving very slowly. More generally, it correctly treats the electrons and photons as quantum particles in any state of motion. It works, in a rough way, as follows. The first electron emits a photon as determined by gauge invariance, and the electron accelerates or recoils. The photon emission represents a mathematical factor of the electric charge, q_a, in the T-matrix. The emitted photon *propagates* to the other electron with a factor that is k/R. The second electron then *absorbs* the photon with an emission vertex factor of q_b. Putting it all together, the overall T-matrix for this process is $q_a \times (k/R) \times q_b = kq_aq_b/R$, which reproduces Coulomb's potential energy. We say that the Coulomb potential, hence the force, arises from the "exchange" of a photon between the electrons.

This is a grossly simplified example of how Feynman diagrams work in practice. With the full machinery of Feynman diagrams, however, we can compute the scattering rate for two beams of electrons colliding with one another, and the experimentalist can compare them with results measured in the lab. Voila! They agree. Of course, we have shed all the technical details here, such as the electron spin, but please accept it on faith—it all works.

As we have seen, special relativity and the quantum theory, when put together, predicted the existence of antimatter. Paul Dirac made this stunning theoretical discovery when he solved the problem of negative-energy electrons in relativity. Carl Anderson then observed the positron,

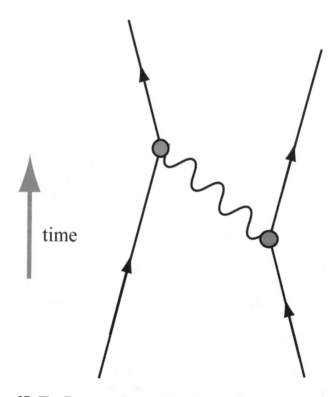

Figure 27. The Feynman diagram for electron-electron scattering. The force between the electrons that produces the interaction arises by the *exchange* of the photon between them. The photon is emitted from the electron as it accelerates, as we described above, as arising from the gauge symmetry of the photon interacting with the electron.

the antiparticle of the electron, a few years later. Feynman, however, reinterpreted antimatter in the new language of his diagrams and gave us a remarkable alternative view as to what antimatter is.

To see this, we look at figure 28, which shows what could be described as an *annihilation* of an electron and positron into a photon, which then makes a *top quark* and an *anti–top quark*. This process could occur in a high-energy electron collider, and a variation on it occurs at the Fermilab Tevatron today, where the initial particles are a quark and an antiquark.

Yet, from another point of view, here we see the electron, with positive energy, approaching the event in space and time at which the energetic photon will be emitted. As this happens, the electron accelerates so

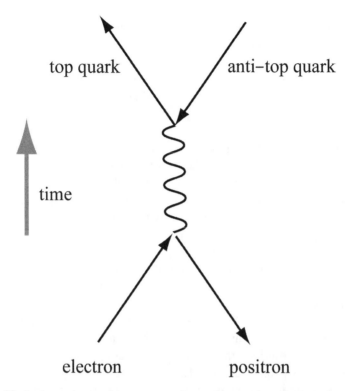

Figure 28. An incoming positive-energy electron (matter) produces a photon and turns around, heading *backward in time* with negative energy. We observe this as the collision of an electron with an antielectron, annihilating into a photon. The photon collides with a negative-energy top quark coming in from the future (an anti–top quark), producing a positive-energy top quark heading back into the future. We observe the electron colliding with the positron to produce the top quark and anti–top quark, each having positive energy.

much that it actually acquires negative energy, and it turns around and starts going *backward in time!* Indeed, Feynman visualized antimatter in this way: antimatter is negative-energy matter going backward in time! Similarly, the emitted photon collides with a negative-energy top quark, coming in from the future, which acquires positive energy as it accelerates. The top quark emerges, heading back into the future as a positive-energy particle! This is a radical reinterpretation of Dirac's idea of antimatter as holes in the negative-energy vacuum.

From Feynman's point of view, this is why antimatter is required to ensure that nothing can travel faster than the speed of light in the quantum world with special relativity. If we forgot to include the backward-in-time electrons, the antielectrons, we would find that signals can then propagate instantaneously from one point in space to another. The emission of a particle wave at time $t = 0$ could, in principle, be detected at the distant star Alpha Centauri instantaneously (this is experimentally untenable). But if we include the negative-energy waves that travel backward in time, we find that they exactly cancel the signal that would have traveled faster than light. If particles had even slightly different properties than antiparticles, such as slightly different masses or electric charges or spins, then the cancellation would fail to be exact, and signals would travel faster than light—and CPT symmetry would be violated!

The real power of Feynman diagrams is that we can systematically compute physical processes in relativistic quantum theories to a high degree of precision. This comes from what we call the *quantum corrections*, or the so-called *higher-order processes*. In figure 29, we show the second-order quantum corrections to the scattering problem of two electrons. This is a set of diagrams, each of which must be computed in detail and then added together including the previous diagram in figure 27, to get the final total result for the T-matrix. (The T-matrix, as noted above, is essentially the quantum version of the potential energy between the electrons and describes the scattering process.) This gives the total T-matrix to a precision of about 1/10,000. We can then go to the third order of higher quantum corrections to try to get even more precise agreement with experiment. Third-order calculations, however, are extremely difficult and very tiring for theoretical physicists. Only the brave and most energetic try them.

With each order of complexity of Feynman diagrams, we have more photon emissions, that is, more electrons emitting or absorbing photons and propagating lines of electrons and photons. The size scale, or "order of magnitude," of a given correction to the basic process is governed by the number of vertices. Each vertex gives a factor of the electric charge, e, but each scattering diagram has at least two vertices, so the series progresses in powers of $\alpha = e^2/4\pi\hbar c$. This particular combination of the fundamental constants is called a "dimensionless number"—all of the physical units (meters, seconds, kilograms) have cancelled out, and we are left with a pure mathematical number with the value of 1/137. Now, this is (thankfully) a small number, so each additional set of Feynman diagrams improves the precision of a calculation of the T-matrix by a factor of only about 1/100. Feynman diagrams to three-loop order have been computed,

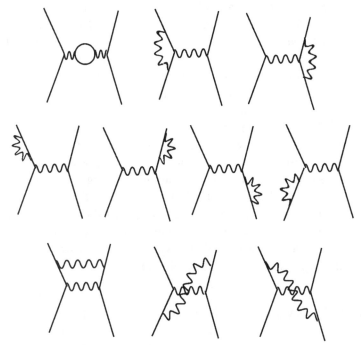

Figure 29. The Feynman diagrams that represent the first order of quantum corrections to the result of figure 27.

and the QED has been checked by experiment to a precision of about one part in 10^{12}. The agreement is perfect. No other physical theory has ever been so precisely tested.

In the second-order processes of figure 29 we now see the appearance of the "loop diagrams." The first diagram contains a loop representing a particle and antiparticle being spontaneously produced and then reannihilating. They contain a looping flow of the particle's momentum and energy. Here we must sum up all possible momenta and energies that can occur in the loops, such that the overall incoming and outgoing energy and momentum is conserved. The Feynman loops present us with a new problem that has bothered physicists in many different ways for many years: put simply, when we compute the loop sums for certain loop diagrams, we get infinity! The processes we compute seem to become nonsensical. The theory seems to crash and burn.

However, as the loop momenta become larger and larger, the loop is physically occupying a smaller and smaller volume of space and time, by the quantum inverse relationship between wavelength (size) and momentum. So in fact, we can only sum up the loop momenta to some large scale, or correspondingly, down to some small distance scale of space, for which we still trust the structure of our theory. At higher loop momentum and energies, more and more uncertainty will enter, because we are probing forces at shorter and shorter distances. At such scales there may be new and different kinds of unaccounted-for phenomena.

Interpreted properly, the loop diagrams actually tell us how to examine the physics at different distance scales, as though we have a theoretical microscope with a variable magnification power. If we carefully measure the electron mass and electric charge at a known scale of distance, then we can reliably predict them at higher energies, or shorter distances. The theory will be completely predictive of all possible experimental measurements up to a certain high-energy, or short-distance, scale. At that energy we must switch over to some other theory such as, perhaps, string theory, and to test those predictions experimentally we need a much bigger microscope, that is, particle accelerator.

TOWARD UNIFYING ALL FORCES IN NATURE

The modern era of gauge theories began with a remarkable paper of Chen Ning Yang and Robert Mills in 1954. These authors asked a straightforward question: "What happens if we replace the gauge symmetry of the electron by another symmetry?" The gauge symmetry of the electron is just the rotation of our dial on the Acme detector; that is, it is just the symmetry of the circle, called $U(1)$. Yang and Mills turned to the next symmetry in the sequence of complexity, $SU(2)$, the symmetry of the sphere in three real dimensions (or the symmetry of the circle in two complex dimensions, or equivalently, the symmetry of a normal sphere in three real dimensions of space; see the appendix). It turns out that this symmetry leads to a more general form of electrodynamics, called a Yang-Mills theory. $SU(2)$ has three gauge fields, hence three photonlike objects, and now the gauge fields themselves carry charges, unlike the case of electrodynamics, in which the photon carries no electric charge. Moreover, the Yang-Mills construction works for any symmetry. Symmetry thus becomes part and parcel of the basic structure of a quantum theory of forces. Little did the world of physics know, at that moment, the door to unifying all forces in nature into one master theory had been opened.

We now know that all known forces in nature are based upon local gauge symmetry theories. This represents a major step toward a unified description of everything in physics. Yet there are four completely different structures, or *styles*, of gauge invariance found in nature. Einstein's theory of gravity contains a coordinate-system invariance; that is, it doesn't matter what coordinate system you use, or how you choose to move, inertially or noninertially through space and time, to describe nature. This leads to gravity as a bending and reshaping of geometry, governed by the presence of energy, momentum, and matter. Particles must then emit and absorb *gravitons*, which are the gauge fields, or the "quanta," of gravity. The Newtonian gravitational theory is recovered only as an approximation at low energies (slow systems, without too much mass).

The description of the remaining nongravitational forces in nature is indeed based upon the Yang-Mills theories. We've just seen how electrodynamics works. However, we've also previously encountered the weak forces, in the explosions of Titans, and we'll see that they are also described by gauge symmetries. These weak gauge symmetries actually flip one species of particle, such as the electron, into another, such as the neutrino. The weak forces are unified with electromagnetism, and they are also intimately associated with the origin of mass of all the elementary particles found in nature.

The strong force holds the atomic nucleus of the atom together, and we'll see that it involves a Yang-Mills gauge field interaction among the particles called quarks. Just as electrodynamics brings the wave function of the electron into an intimate contact with the photon, the weak and strong forces are intimately interwoven with the detailed patterns and properties of the elementary particles. In fact, the distinction between "elementary particle" and "force" almost becomes artificial in modern physics. But what is an elementary particle? To this question we must now turn our attention.

chapter 12

QUARKS AND LEPTONS

Who ordered that?
　　　　　—I. I. Rabi, on hearing of the discovery of the muon

O ver the past several centuries, people have come to believe in *atoms* on the basis of the compelling evidence from chemistry. Atoms were thought to be the basic "elements" that would not change their properties under chemical reactions. Alchemists could never succeed in turning the element lead (Pb) into the element gold (Au). Through countless attempts to do so, they merely rearranged elements within many substances and amassed an enormous database that formed the foundation of the science of chemistry.

Matter, at the first stage of deconstructing it, is largely found in the form of *molecules*, which are either large or small assemblages of atoms. "Elements" are essentially synonymous with "atoms." Salt, for example, is a molecule, a combination of the atom (element) sodium with the atom (element) chlorine; water is a molecular combination of two atoms of hydrogen and one atom of oxygen; methane is a molecule containing four atoms of hydrogen and one carbon atom. And so forth. Thus salt, water, and methane are molecules that can, in principle, be broken down chemically into their constituent atoms—with enough effort in the alchemist's lab. But there it ends for chemistry. Sodium, chlorine, hydrogen, oxygen, carbon, and so on are the *fundamental particles* of chemistry. They cannot be further subdivided without initiating much higher-energy processes than are possible in a chemistry lab.

By the mid-nineteenth century, the elements were classified according to their chemical properties by Dmitry I. Mendeleyev. This led to the familiar *periodic table of the elements*, which hangs on the wall of every high school chemistry classroom. The columns of the table represent elements with similar chemical properties. The table is "periodic" because it repeats itself in a pattern that was a puzzle to nineteenth-century scientists. It awaited the invention of the quantum theory for its elucidation. The periodic table of the elements was, nonetheless, a pinnacle summary of hundreds of years of alchemy and science, representing the reduction of the virtual infinity of molecules into approximately one hundred fundamental atoms found in nature. (Many of the heaviest elements have only recently been artificially produced and are very short-lived; they are usually not to be found on our old high school classroom walls.) The periodic table represented a pattern of complexity in the properties and forms of atoms, and it suggested that atoms themselves must have internal structure and that a deeper layer of *subatomic* matter must exist.[1]

The detailed understanding of the atom began about fifty years after Mendeleyev, with Thompson's discovery of the electron, Rutherford's discovery of the nucleus, and Bohr's rudimentary theory of the electron orbits, based upon the newborn quantum mechanics. Atoms are indeed made of smaller objects. We thus traversed, from Mendeleyev to Bohr, from molecules to atoms, to finding even more elementary objects inside the atom—the nucleus and the electron, and then the proton and neutron within the nucleus. It was as if a sequence of Russian dolls was being opened, the last of which always revealed yet another Russian doll inside. Where would it end? Perhaps the things inside the atom would be the last and tiniest of the Russian dolls. The tools were in place to dissect matter into its most fundamental parts, the tools being special relativity and quantum mechanics. Thus began the science of elementary particle physics—the most fundamental science of all.

INSIDE THE ATOM OF THE MID-TWENTIETH CENTURY

By the early twentieth century scientists understood that an atom is much like a solar system (see fig. 30). At the center is the analogue of the Sun: the nucleus of the atom. The nucleus itself is a composite object, containing protons and neutrons. Any particular atomic element is defined by the *number of protons* in its nucleus (which is equivalent to the nucleus's electric charge). For example, the nucleus of hydrogen contains a single lone proton, whereas a nucleus of an atom of carbon always contains six protons. In addition to the protons in the nucleus, we find the electrically neutral (uncharged) particles we call neutrons. The number of neutrons in a nucleus of an atom can vary for a fixed number of protons. Hence, carbon-12 has 6 protons and 6 neutrons, whereas carbon-13 has 6 protons and 7 neutrons, and so forth. These different carbon nuclei with a proton number of 6 but different neutron numbers are called the *isotopes* of carbon.

The atomic nucleus is held together by a very strong force—which is, in fact, called the *strong force*. This force has to be strong because the protons each have positive electric charges, and therefore repel one another electrically. The nucleus would fly apart unless an overwhelmingly strong force compensated for this repulsion and bound the protons, together with the neutrons, into the compact nucleus. The strong force was found to arise from other particles called *pions* (or π-mesons), which hop back and forth between the protons and neutrons (just as photons, the particle of light, create the electric force by hopping between electrically charged particles in a Feynman diagram). The nucleus of an atom is indeed very dense and compact, typically having the size scale of 10^{-15} meters. In any given atom, 99.95 percent of the mass resides in the nucleus.

Orbiting the atom at comparatively larger distances, the analogues of the planets orbiting the Sun are the electrons. The orbits of the electrons typically have a size of 10^{-10} meters, and the electrons are bound to the atom by electric forces, through the attraction of their negative electric charges to the proton's positive charges. In its normal, electrically neutral state, the electrons balance the number of protons of the atom. The electrons do not feel the strong force; their motion is governed by the laws of quantum mechanics, and their orbits form cloudlike configurations.

The sharing, or "hopping back and forth," of outer-electron orbital motion between two atoms leads to the forces that bind atoms together to

Electrons move in
quantum orbits around the
nucleus, bound by the electric
force.

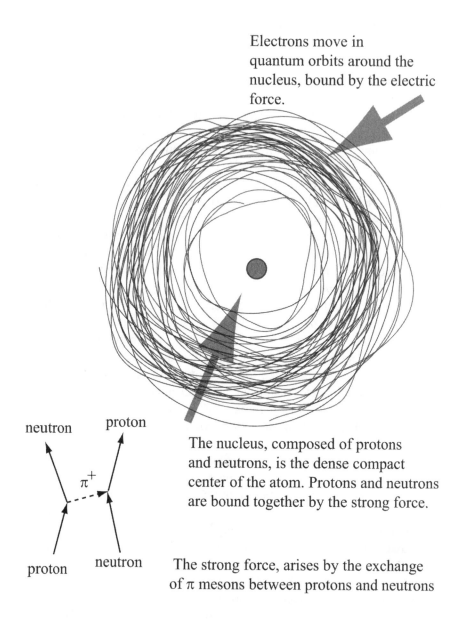

neutron   proton

π^+

proton   neutron

The nucleus, composed of protons
and neutrons, is the dense compact
center of the atom. Protons and neutrons
are bound together by the strong force.

The strong force, arises by the exchange
of π mesons between protons and neutrons

Figure 30. A schematic view of the atom.

become molecules. The details of this force are somewhat complex, and a large variety of atomic combinations, hence molecules, are possible. As we ascend back up the chain to molecules, this increasing complexity leads to the richness of our world, much like ascending from a box of oil paints, to the vast variety of masterpieces found in the world's art museums. Thus all of the vast array of chemical phenomena is explained in terms of the quantum motion of electrons, with their electromagnetic forces, resulting from the quantum-mechanical exchange of photons between them—which is a consequence of the gauge symmetry.

Indeed, one of the profound lessons of the twentieth century, what we have seen in the previous chapter with Feynman diagrams, is that "force" and "particle" blend together to become a common unified entity. Forces result from the *exchange of particles* (such as photons) between other particles (such as electrically charged electrons or protons)—like the back-and-forth pattern found in the music of Bach, underlying and subtly defining the grand musical composition that is nature.

To summarize the situation, by the early to mid-twentieth century, all of the known particles, protons, neutrons, and pions that make up the atomic nucleus were thought to be pointlike and elementary. Pions were theoretically predicted by Hideki Yukawa in 1935, based upon the known properties of the atomic nucleus and the requirement of the contemporary theory for a new particle that could hop back and forth between protons and neutrons, thus accounting for the strong force. A particle called the *muon* was suddenly, and coincidentally, discovered in 1937, through cosmic-ray observations, with nearly the same predicted mass as the pions. This initially caused great confusion, as the muon was initially thought to be the pion, but it did not interact strongly with protons and neutrons, so it could not be the agent of the strong force, as predicted by Yukawa. In fact, the muon seemed to be a mere carbon copy of the electron but two hundred times heavier (and decaying in a millionth of a second). Later, however, the pions were discovered, vindicating Yukawa's ingenious theory, for which he won the Nobel Prize. The muon seemed to be a fluke, and its appearance elicited I. I. Rabi's famous quip, "Who ordered that?" But science was on the verge of opening yet another Russian doll.

We must note that we are now about to leave the realm of normal, everyday physics and enter the world of the elementary particles. In this world, the currency, or units of measure, especially the kilogram, become awkwardly inconvenient. To quote the masses of elementary particles we use Einstein's famous equation, $E = mc^2$, and we thus use energy as a measure of mass. The convenient unit of energy is the *electron volt*. An electron volt is the amount of energy that a one-volt battery expends when it pushes an electron through an electric circuit. This is a tiny amount of energy, because an electric circuit normally involves many trillions of electrons passing through it. However, electron volts provide a convenient system of units for quoting the masses of elementary particles. The electron has a mass in these units of 0.511 *million* electron volts, or MeV. The proton is much heavier, with a mass of 0.938 *billion* electron volts, or GeV (giga-electron volts).[2]

QUARKS

In the early 1950s an array of new and unanticipated particles was being produced by colliding together protons with the nuclei of atoms, using the burgeoning technology of energetic particle accelerators. The list of new particle discoveries grew rapidly and soon exceeded the number of atomic elements. There was emerging a morass of "fundamental" particles. All of these various new particles were strongly interacting cousins of the proton and neutron, and the pion, the components of the atomic nucleus. They were unstable, having extremely short lifetimes, and thus could not make up ordinary matter found on Earth. As these new strongly interacting particles proliferated, only one tool could be brought to bear to try to make sense of them: symmetry.

No physicist of the era was more a master in implementing the tool of symmetry than Murray Gell-Mann.[3] Gell-Mann, a child prodigy, made his first significant physics contributions in his early twenties. He recognized early on that symmetry was an essential tool, that it led to classification schemes, to relationships between properties, and then to successful predictions of the quantitative properties of the particles. Gell-Mann, like Mendeleyev a century earlier, identified the key patterns in the emerging zoo of strongly interacting particles, using the sophisticated mathematics of symmetry groups, in many ways teaching the rest of the community of physicists how to think in the arcane language of quantum symmetries.

There had emerged complexity in the vast assortment of the strongly interacting particles that indicated that these were not fundamental. The symmetries of these particles, like the recurring chemical properties of atoms, hinted at the existence of yet another layer. Yet there were serious problems with this idea of another stratum of nature—whatever comprised the strongly interacting particles could never be set free. Even the most powerful of particle accelerators, producing the most violent collisions, never liberated any of the inner components, but rather simply produced more and more of the unstable strongly interacting particles. Nonetheless, for the hypothetical next layer of constituency of matter, whether real or purely mathematical, Gell-Mann introduced the term *quarks*, borrowed from James Joyce.[4] Finally, in the early 1970s, the first "high-resolution photograph" of the inner world of the proton was taken at the Stanford Linear Accelerator, and for the first time, quarklike structures were seen. The strongly interacting particles were thus unraveled, by the concept of symmetry and experiment, to reveal another Russian doll, the elementary constituents of the strongly interacting particles—the quarks. Quarks, in fact, do exist in nature, and their properties can be measured. Yet, enigmatically, quarks can never be set free from the prisons of the strongly interacting particles that they compose.

The discovery of the quarks is a fascinating and heroic story, but it is also a long one. So let us fast-forward to the present and survey what is now known about the basic building blocks of matter.

THE STANDARD MODEL OF PARTICLES AND FORCES

A graduate student beginning her studies in elementary particle physics, as in zoology, is overwhelmed by the vast array of species and nomenclature. Though zoology has a multitude of species, an overarching classification system of creatures exists, thankfully owing to patterns in the species that emerged through the process of evolution. Once the budding zoologist knows the difference between the phylum of worms called Acanthocephalia (spiny-headed, parasitic worms, with about 1,150 species) and the phylum called Nematoda (roundworms, with about 12,000 known species), then she doesn't have to get too bogged down with the details of any particular subspecies, unless that is her chosen field of specialization.

For particle physics there is a simpler classification scheme and fewer creatures; yet it can all be a bit daunting on the first viewing. This pattern of particles comes from the laws of physics, but we don't yet know how or why. It is a puzzle, like Mendeleyev's periodic table of the elements was before the advent of the quantum theory. The pattern of the elementary particles is also periodic. Within the array of the species of quarks and leptons, and the gauge bosons, we see patterns and apparent symmetries—yet no Niels Bohr has yet arrived to explain it all in a predictive way. Perhaps the young graduate student entering particle physics, through diligent studies and fresh imagination, will succeed.

The things we call "elementary particles" today, to the best of our present knowledge, are structureless, pinpoint shards of matter. All experimental data through 2004 indicates that these particles, while rich in their diverse properties, have physical internal dimensions, or sizes, of zero! Like Alice's Cheshire cat, the particles can be imagined to shrink to zero size, leaving behind only a smile with their other various properties like spin, charge, mass, and so on.

For the elementary particles, there are essentially three "phyla," two containing the basic building blocks of matter, *quarks* and *leptons* (see table 1), and a third, which contains the *gauge bosons*, the particles that we generally say are responsible for the forces in nature (see table 2). Fortunately, the phyla of the known particles are fairly simple and much smaller than the phyla of life on Earth.

The so-called "matter particles" are the quarks and leptons. Each of these particles is like a tiny gyroscope and has a spin of 1/2, in accordance with the rules of quantum mechanics. All of the everyday matter in our world is composed of the two quarks called *up* and *down*, as well as one lepton, the electron.

These objects are distinguished by their electric charges and their masses. We always define the electron to have an electric charge of -1. In these units, the up quark (u) has an electric charge of $+2/3$, and the down quark (d) an electric charge of $-1/3$. The proton is therefore not an elementary particle but rather a composite particle, built of three quarks in the pattern $u + u + d$ (or uud). Adding up the electric charges of the constituent quarks, we see that the proton charge is $+2/3 + 2/3 - 1/3 = +1$. Similarly, the neutron is composed of $u + d + d$, and the corresponding electric charge combination is $+2/3 - 1/3 - 1/3 = 0$.

As we have seen, Einstein's special theory of relativity, combined with quantum mechanics, requires that every particle in nature have a corresponding antiparticle. All of the naturally occurring antimatter has now disappeared from our universe, for mysterious and uncertain reasons, but it can be re-created in the lab, and it exists for fleeting instants of time inside the strongly interacting particles themselves. The antiquarks have

Quarks   Leptons

charge		mass	red	blue	yellow		mass	charge

First Generation

charge		mass	red	blue	yellow		mass	charge
+2/3	up	0.005 GeV	u	u	u	• electron neutrino		0
−1/3	down	0.01 GeV	d	d	d	• electron	0.005 GeV	−1

Second Generation

+2/3	charm	1.5 GeV	c	c	c	• muon neutrino		0
−1/3	strange	0.15 GeV	s	s	s	• muon	0.10 GeV	−1

Third Generation

+2/3	top	178 GeV	t	t	t	• tau neutrino		0
−1/3	beauty	5 GeV	b	b	b	• tau	1.5 GeV	−1

Table 1. A periodic table of the quarks and leptons. In addition, there are the antiparticles, required by the symmetry of special relativity. Antiparticles have opposite electric charges and anticolors; hence, the blue quark has an antiparticle that is "antiblue," which acts like a combination of red and yellow. The neutrinos have extremely tiny masses, expected to be less than or of the order of 1 electron volt. The neutrino masses and their effects (called *oscillations*) are recent discoveries and are presently a very active area of research in elementary-particle physics.

Gauge Bosons

charge	mass		mass
Electroweak		**Strong (gluons)**	
0 photon	0 GeV	(red, antiblue)	0 GeV
+1 W^+	80.4 GeV	(red, antiyellow)	0 GeV
-1 W^-	80.4 GeV	(blue, antired)	0 GeV
0 Z^0	90.1 GeV	(blue, antiyellow)	0 GeV
		(yellow, antired)	0 GeV
Gravity		(yellow, antiblue)	0 GeV
		(red, antired) – (blue, antiblue)	0 GeV
0 graviton	0 GeV	(red, antired) + (blue, antiblue) – 2 (yellow, antiyellow)	0 GeV

Table 2. Table of the gauge bosons. These are also the known "force carriers," and all are defined by gauge symmetries.

the opposite electric charges to their quark counterparts. We designate the anti–up quark as $\bar{u}$, and it has an electric charge of –2/3, whereas the anti–down quark is $\bar{d}$, with an electric charge of +1/3. The pions, the particles that hold protons and neutrons in the nucleus, are composed of combinations of a quark and antiquark. We easily see that there are four possible quark-antiquark combinations involving u, d, $\bar{u}$, and $\bar{d}$: $u\bar{d}$, (–1), $\bar{u}u$ (0), $\bar{d}u$ (+1), $\bar{d}d$ (0). In quantum mechanics, neutral-particle wave functions often become "blended" (i.e., added together in particular ways), and the resulting composite particles that we observe in the laboratory are $\pi^+ \leftrightarrow \bar{d}u$, $\pi^0 \leftrightarrow \bar{u}u - \bar{d}d$, $\pi^- \leftrightarrow u\bar{d}$, and $\eta^0 \leftrightarrow \bar{u}u + \bar{d}d$. The first three are the pions, and the fourth is called the "eta-meson." All four are known well from experiment, and their quark composition accounts neatly for the pattern. In fact, from the masses of the pions and other mesons, we can deduce the masses of quarks themselves, as indicated in table 1.

Atoms also contain electrons. The electron is also a truly fundamental particle, and it is known as a lepton. We note that most newspaper articles about elementary-particle physics usually make the statement that all matter is composed of quarks. This is untrue, since leptons are not composed of quarks or anything else that we can see; they are themselves elementary particles. The quarks lay low within matter, making up the protons and neutrons and pions, deep inside the nucleus of the atom. From the viewpoint of chemistry, the nuclei of atoms are simply ballast. It is the dance of the electrons, as they orbit and hop about the atoms, that leads to the diversity of the chemical and biological world.

We have also seen that the supernova explosion of a Titan is driven by the process $p^+ + e^- \rightarrow n^0 + \nu_e$, which involves yet another leptonic particle, the electron neutrino, ν_e. Indeed, processes like this are occurring at this very instant deep inside the core of the Sun (don't worry, our Sun is not about to go supernova). Billions of electron neutrinos are streaming out from the Sun and penetrating our bodies every second. The neutrino has a zero value of electric charge, as well as a very tiny and almost negligible mass. Neutrinos have no electrical interactions or strong interactions (thus they are leptons); hence, they interact very weakly with the rest of matter.

Taken together, the quarks u and d and the leptons e and ν_e constitute a "family" that we call the *first generation*. These are the lightest, in mass, of the quarks and (charged) leptons. The generations of quarks and leptons form a pattern, as seen in table 1. We should emphasize immediately that we have no deeper understanding, though many theories abound, as to what we really mean by the "first generation." It is a convenient representation but not an established scientific one, yet. We are, after all, approaching the frontier of our knowledge about the world, and, like a battlefield, things are becoming tentative—the ground can shift out from under our understanding at any moment!

So, what dictates the structure of a given generation? First, these four particles are the lightest of their types, and we therefore group them together on the basis of mass, hoping that some symmetry will someday explain this grouping in greater detail. Furthermore, we notice that, counting every particle in a generation, including each of the quark colors, the net electric charge of everything adds to zero. That is, the charges of three up quarks plus three down quarks, plus the electron and the neutrino is $(3 \times 2/3) + (3 \times -1/3) - 1 = 0$. Zero! This is further evidence of a pattern, and it suggests deeper symmetries. However, we don't yet know the exact origin of this pattern.[5]

In many ways, we might have stopped here if the design of the universe was up to us. All the matter and relevant everyday processes in nature seem to involve only the four objects of the first generation. There seems at present to us to be little practical need, requirement, or benefit of anything else in nature. We therefore do not understand the mind of nature, because nature mysteriously provides us with another two generations of quarks and leptons in exactly the same pattern with essentially the same properties, but more massive.[6]

The second generation, listed in table 1 above, contains the quarks c, for *charm*, and s, for *strange*, as well as the leptons called the *muon* and ν_μ. Even at the outset of their discovery, these particles seemed to be pointless addenda to the list of ingredients to the physical world (Rabi's famous quip again comes to mind, "Who ordered them?") And if the second generation seemed pointless, the third generation seems wholly unnecessary, containing the *top* and *beauty* quarks as well as the additional leptons *tau* and ν_τ. We thus see three complete generations in nature of quarks and leptons. Each successive generation is a copy of the previous one, only heavier in mass. Why the pattern within a generation? Are there only three generations? What dictates the pattern of masses of each generation? These are open questions. They beg further experiments, for theorists aren't helping much to answer them.

There are indications, however, that the top quark is truly the top, and the end of the sequence. We now have indirect experimental indications, from detailed studies of the *weak interactions*, that there are no more generations of quarks and leptons—at least not in this particular pattern.[7] Furthermore, the top quark mass is enormous on the scale of the other quarks and leptons, and indirect evidence suggests that there is no room remaining for yet another heavy quark. In fact, the present situation tantalizingly suggests that we may be getting close to answering one of nature's basic questions: where the masses of the elementary particles come from. The top quark may be playing an intimate role in this, or at least it is the spectator with the best front-row seat. To appreciate this issue, we must turn to the forces of nature.

Indeed, something must be binding the quarks and antiquarks together inside of the strongly interacting particles—the proton, neutron, pions, and the long list of associated creatures found in the 1950s and 1960s. This is related to the *strong force*, by which the resulting composite particles, protons, neutrons, and pions interact. But this must be the strong force acting at the next more fundamental layer, acting among the quarks themselves. The strong force among quarks leads to a complexity of its own.[8]

Only special combinations of quark composites are observed experimentally to occur. In nature we only find objects containing three quarks, called *baryons* (or three antiquarks, called antibaryons), or objects containing a quark and an antiquark, called *mesons*. Whatever the strong force is at the quark level, it has to account for this pattern. So the question arises, what is the nature of the strong force that holds the quarks together inside the hadrons? In fact, as we have mentioned, there have been countless attempts to try to liberate quarks in experiment, yet we find that quarks are always imprisoned inside the hadrons they comprise. This issue turns on a fundamental and subtle property of the quarks, which in turn reveals a new symmetry of nature.

Examining the quarks in greater detail in table 1, we find that they each come in "triplets." That is, there are three types of up quark, three types of down quark, and so on. We call this additional label on the quarks *quark color*. Hence, we say that there is a "red up," a "blue up," and a "yellow up quark." This has nothing to do with visual colors of the rainbow but is rather a fanciful and mnemonic description of the full symmetry of quarks.

The color of a quark is physically hard to detect, because any observed particle that the quarks comprise, by definition a hadron, always has a net color of *zero*. For example, at any instant of time the proton contains *uud*, but one quark is red, another blue, and another yellow, making an overall color-neutral state.

The antiquarks must be viewed as having anticolors in the sense of the color wheel. So the anti–blue up quark is actually a red-yellow, or "orange," object. Therefore we can make color-balanced mesons by combining pairs of quark and antiquark. This simple rule explains the forms of the bound particles that we see; however, it also gives the clue to the fundamental theory of the strong interactions.

THE STRONG FORCE IS A GAUGE SYMMETRY

How do we know that quark color exists if we can't see it? In fact, it was anticipated in the early days of the quark theory because of the exchange symmetry of identical particles. There exists a composite strongly interacting particle whose properties Gell-Mann dramatically and precisely predicted in 1963. Experimentalists quickly confirmed the prediction at Brookhaven National Laboratory. This is the Ω^-, the "omega-minus" particle, and it contains three strange quarks, or *sss*. It is known that the quarks making up the Ω^- must move in a single common orbital, but without quark color this would be *strictly forbidden* by the exchange symmetry; that is, it would represent three identical fermions in the same quantum state (see chapter 10). Yet the Ω^- does exist. The only way out of this conundrum is the existence of quark color. If one s-quark is "red," the second "blue," and the third "yellow," making up the Ω^-, then the three quarks are not identical and there is no problem with all three occupying the same quantum state of motion at the same time. There are many other ways in which the number of colors of quarks have been "counted" in experiment, and the result is always consistent with three.

This leads us to the question of the true nature of the quark color symmetry. We can think of quarks as though they live in a three-dimensional space, whose three axes are labeled by the three colors. In this space, a quark can be thought of as an arrow (a vector) that can point in any direction. If the quark is red, its arrow points along the red axis; if blue, then the blue axis; and so forth—but the arrow can rotate and point in any direction. The color symmetry is just the collection of rotations that we can do to such a quark arrow (this is the symmetry group $SU(3)$; see the appendix).

Now suppose we generalize the idea of the previous chapter, the idea of gauge invariance. Gauge invariance means that we can change the unobservable "phase" of the wave function of a quark (the pointing direction of the arrow on an Acme wave function detector), as we did for the electron. This change scrambled the electron's energy and momentum. For the electron, the price we paid for the symmetry was the introduction of the photon, to "undo" the scrambling change we make (to counterrotate the dial on the Acme detector) and restore the original electron's momentum and energy. The electron becomes a blend of its own wave function with the gauge field. By shaking the electron, that is, by causing it to accelerate, we can cause the emission of a physical gauge particle, or *gauge boson*, called the photon.

But for quarks, let us now generalize the symmetry concept. Let us allow the change (gauge transformation) that we do to the quark's wave function simultaneously to be a *rotation* in the color space, together with a change in the quark's *momentum* and *energy*. Therefore we can now perform a transformation that, for example, rotates a pure red down quark into a blue down quark, as we simultaneously scramble the momentum and energy of the quark. We want this to be a symmetry, and we want to end up in a state that still has pure red color, as well as the same energy and momentum that we started with. Just as we saw previously for the electron, this requires that we have additional particles that "undo" the changes we make on the red quark, unscrambling the energy and momentum, keeping the overall result invariant.

To have such a color gauge symmetry, we need eight new gauge particles, called *gluons*. Gluons are emitted from the quarks like photons, but they carry off the old color of the quark and carry in a new color. So a gluon has a color and an anticolor. When we thus do a local gauge rotation, starting with a red quark and turning it into a blue quark, as in our example, then we simultaneously create a (red, antiblue) gluon, so the net color is red + antiblue + blue = red, and therefore initial color of the quark, red, is recovered. The gluon also compensates for the scrambled momentum and energy information, and the final quark quantum state has the same momentum and energy as we started with (see fig. 31). We thus have a new gauge symmetry, and a new force of nature, associated with quark color![9] The experimental evidence favoring the existence of gluons, gathered since the 1980s, is now very solid.

If the quark accelerates, then a physical gluon is emitted. Note that the emission of a (red, antiblue) gluon can occur from a red quark, and the quark is now turned into a blue quark. And, if a gluon collides with a quark, the gluon will be absorbed and the quark will be accelerated. It is, perhaps, one of the most astounding aspects of modern science that the simple idea of a symmetry, local gauge invariance, which yields up the photon and quantum electrodynamics, likewise yields the correct theory of the strong interactions when adapted to quark color. This theory is called *quantum chromodynamics* (QCD for short) and it, too, is a stunning success.

Quarks thus interact with one another by the exchange of gluons (see fig. 32). We can draw the appropriate Feynman diagrams, and we learn how to compute them. The force is strong because the "color charge," the analogue of the electric charge, is large.

The emission of a gluon
from a quark

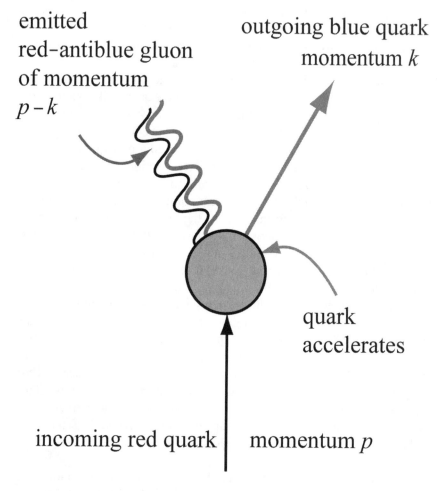

emitted
red-antiblue gluon
of momentum
$p-k$

outgoing blue quark
momentum k

quark
accelerates

incoming red quark | momentum p

Figure 31. A quark accelerates and changes color from red to blue, and emits a (red, antiblue) gluon, so the total color is preserved. The momentum and energy are also conserved.

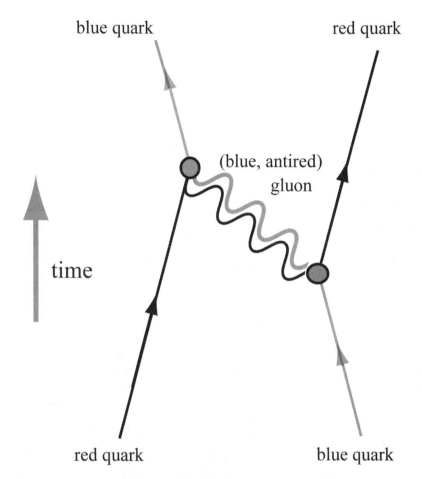

Figure 32. A red quark scatters off a blue quark. The quarks exchange color through the (red, antiblue) gluon that hops between them, giving rise to a strong force. The proton is held together by gluon exchange among the quarks. Every 10^{-24} seconds or so, a gluon hops between quarks in the proton.

One of the remarkable discoveries about QCD is that the coupling strength of the interaction of the quarks to the gluons, denoted g_3 (as mentioned before, the analogue of the electric charge, e), actually becomes weaker as we push the quarks to extremely short distances. Conversely, at large distances, the quark-gluon coupling strength becomes very great. This leads to a strong pull on the quarks, preventing them from being separated and isolated in the lab. It also turns out that, because of this strong coupling, only quantum-bound states composed of quarks that have an exact color neutrality—a perfect balance of all three quark colors at any instant in time—can exist. This means we can have only combinations of rby, which are the baryons, or $(\bar{r}\bar{b}\bar{y})$, the antibaryons (by $(\bar{q})$ we mean the anticolor of q), or the color-neutral quantum combination of $(r\bar{r} + b\bar{b} + y\bar{y})$, which are the mesons. The color gauge theory, QCD, neatly explains the pattern of strongly interacting particles discovered in the accelerators over the past three decades. It also explains why the quarks can never be really freed from their prisons.

Whereas it is very hard to compute the properties of the theory when g_3 is large, the fact that g_3 becomes small at short distances means that fairly precise calculations using Feynman diagrams can be performed, revealing the collisions and scattering of individual quarks at very high energies. It also means that at very high energies, for example, in the collisions produced at Fermilab's Tevatron (see fig. 33), the individual quarks and gluons collide and leave traces of their collisions. This leads to a spectacular phenomenon, nature's own version of a prison break, known as a *quark jet* (there can also occur gluon jets).

At the Tevatron, a proton with one trillion electron volts (1 TeV) of energy collides head-on with an antiproton of the same energy. The proton contains *uud*, whereas the antiproton contains the three antiquarks $\bar{u}\bar{u}\bar{d}$. At the highest energies, or shortest instants of time, the individual quarks are resolved and behave as though they were almost free particles. Therefore collisions occur in which a pair of quarks, perhaps a *u* and a $\bar{u}$, collide head-on. This quark and antiquark are scattered through very large angles, ripped out of the proton and antiproton, while the remaining debris, the other quarks and gluons of the original proton and antiproton, continue to move forward in their original directions of motion. For a brief moment the quark and antiquark are free, moving at very high energies, hence ultrarelativistic, and they can travel perhaps a hundred times the distance within which they are normally confined, away from their brethren debris of quarks and gluons of the shattered proton and antiproton. The quarks, for a brief moment in time, have broken out of their confining prison cells.

But then the strong interactions take over, and the vacuum itself begins to rip apart in the vicinity of the collision. Pairs of quarks and antiquarks and gluons are conjured forth, ripped from the vacuum itself by the ferocious energy of the collision, a turbulent plasma of matter resembling the initial instant of creation streaks from the point of collision, like

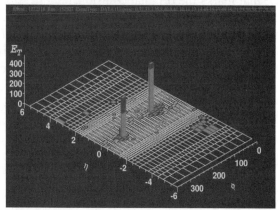

Figure 33. Two massive detectors (CDF and D-Zero) are used at Fermilab to observe proton collisions, head-on, with antiprotons. The beam of protons and the beam of antiprotons pass through the center of the CDF detector, shown here while removed for upgrading. The beams travel in opposite directions at 99.9995 percent of the speed of light. The detector is shaped like an enormous barrel, wrapped around the collision point at the center of the detector. The collisions involve the annihilation of quarks within the proton and antiquarks within the antiproton. The lower picture shows the result of such a collision. Visualize the detector as a tube that is unwrapped into a flat sheet. The squares are "pixels" that register the deposition of energy in the detector. The height of the column, shaped like a stack of Lego blocks, is the energy recorded in that pixel. This shows a collision producing an extremely energetic electron and positron. This is one of the few most energetic collisions ever seen by human beings and probes the structure of space itself at the shortest distances ever examined, less than 1/10,000,000,000,000,000,000 of a meter. (Photos courtesy of Fermilab.)

the long arm of the law, apprehending the escapees. The liberated quarks become shackled by this flurry of new matter and antimatter. Soon all quarks and gluons are captured, reassigned to new pions, protons, and neutrons. The liberation of the quarks is over.

Nonetheless, the indelible footprint of the escaped quarks remains (see fig. 34). Two very well-defined blasts of particles, the jets mentioned above, mostly composed of pions, stream off into space in the directions of the original u and $\bar{u}$ escapees. These jets of particles clearly mark the paths and carry the full energy of the liberated quarks. These jets are the conspicuous tracers of quarks. Thus we can see the structure of the original collision event and the behavior of quarks momentarily celebrating freedom, with our gigantic particle detectors wrapped around the tiny region of space that marks the original collision.

The quarks can also annihilate momentarily into a gluon, which then quickly shreds the vacuum, producing a top quark and an anti–top quark (see fig. 35). The decay signature of the top quarks is reconstructed in the detector. In this way, nature's most massive elementary particle—the top quark—the most recent one to join the "discovered" list, is pulled from the depths of the vacuum sea of buried matter that surrounds us. The top quark is a mighty catch, a pinpoint shard of matter as heavy as a nucleus of a gold atom. The massive monster, the top quark, begs the answer to a question: What gives rise to the masses of the quarks and leptons?

THE WEAK FORCE

We have now described in some detail three of the forces in nature: electromagnetism, the strong "color" force of QCD, and gravity. There remains a final force that defines in a more fundamental way the identities of the particles. This is the *weak force*, and its description as a gauge symmetry unifies with the electromagnetic force. It places us on a road toward the ultimate unification of all forces. Taken together, the quarks, leptons, and the gauge symmetries defining all known forces provides a complete accounting of almost all observed physics to date, and it defines what is called the *Standard Model*.

More than sixty-five years ago Enrico Fermi wrote down the first descriptive quantum theory of the "weak interactions." At that time, these were the feeble forces seen at work in nuclear processes such as beta-decay radioactivity, which, as we have shown earlier, could be considered

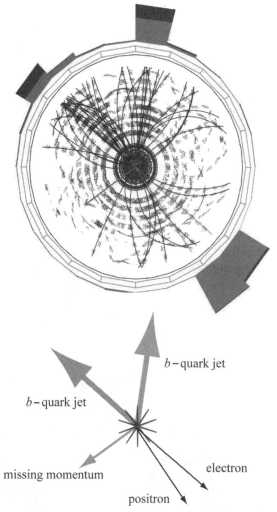

Figure 34. Here we see a collision, looking along the direction of motion of the proton, as it collides head-on with the antiproton, throwing the debris of many elementary particles produced in the collision outward into the detector. The detector has a large magnetic field that bends the tracks of charged particles, enabling their identification. The lower figure shows the main energetic parts of this particular collision's debris. We see two leptons (an electron and a positron) and two well-collimated jets of particles, tracing from a beauty quark and an anti–beauty quark. The event is also missing a large amount of energy and momentum, in the form of outgoing neutrinos. This event can be interpreted as the production of a top and anti–top quark pair, as described in figure 35. (Photo courtesy of Fermilab.)

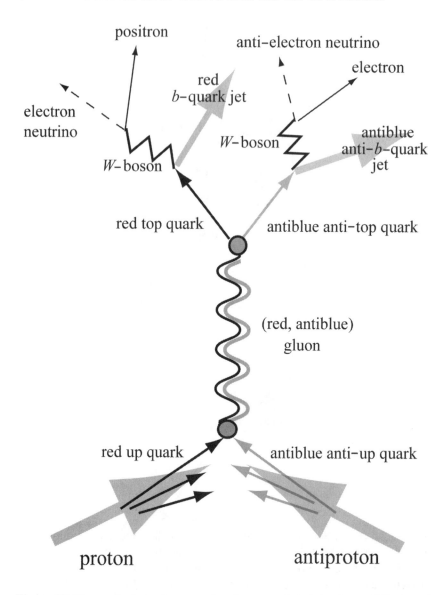

Figure 35. The production of a top and anti–top quark pair by the annihilation of an up quark (from the proton) with an anti–up quark (from the antiproton), through an intermediate gluon. The top quark subsequently decays into a W-boson and a b-quark (which makes one of the jets). The W-boson then decays into a positron and a neutrino. Similarly, the anti–top quark decays into antiparticles. The neutrinos cannot be detected, emerging instead as "missing momentum and energy."

the gunpowder of the supernova. Fermi had to introduce a new fundamental constant into physics to specify the overall strength of the weak interactions, much like Newton had to introduce the gravitational constant G_N. In fact, Fermi's constant is called G_F, and it represents a fundamental unit of mass that sets the scale of the weak forces, which is about 175 GeV.

The weak forces were later found to involve a local gauge symmetry. These theoretical discoveries of the architecture of the Standard Model— led by Sheldon Glashow, Abdus Salam, and Steven Weinberg and perfected as a quantum theory by Gerard 't Hooft and Martinus Veltman— constituted a revolution in particle physics that occurred in the early 1970s, about the time that the quarks were first glimpsed in experiments. This was the decade when it became both theoretically and experimentally established that all forces in nature are governed by the overriding symmetry principle: gauge invariance. We have seen already how this works for the strong color force and electromagnetism.

What, then, is the gauge symmetry of the weak interaction? We see that, within each generation, quarks and leptons are paired. That is, the red up quark is paired with the red down quark, the electron-neutrino with the electron, the charm quark with the strange, the top with the bottom, and so forth. By now, the process of "gauging a symmetry" may seem familiar. We thus imagine that the electron and its neutrino are a single entity that lives in a two-dimensional space, with one axis labeled "electron" and the other "electron-neutrino." The quantum object is an arrow in this space that can point in any direction. When the arrow points along the electron axis, we have an electron. Rotating the arrow, we have a neutrino. The rotations we can do on the arrow form the symmetry group, called $SU(2)$ (as described in the appendix).

So, we now imagine an incoming electron-neutrino wave function with a given momentum and energy. Then we perform a *gauge transformation* that rotates this into an electron, which has negative charge and also scrambles the electron momentum and energy. To make this into a symmetry, we need to introduce a gauge field, called the W^+, which we can simultaneously turn on to restore the total energy and momentum, and rotate the quantum arrow back to its original, electrically neutral "electron-neutrino" direction. In a sense, the gauge field rotates the coordinate axes so that the arrow is now pointing back in the original direction, relative to the coordinate system, and we get back the original neutrino we started with. This is completely analogous to what we've done above with quark color, where the gauge rotation from one color to another is compensated by the gluon field.

It turns out that this requires a total of three new gauge particles, the W^+, W^-, and Z^0, and these are closely related to the photon. In fact, electrodynamics and the weak interactions become blended together, by symmetry, into one combined entity called the "electroweak interactions." At the level of quarks and leptons, the decay of the neutron (by now familiar), if viewed under an extremely powerful microscope, becomes the process of an individual down quark decaying to an up quark, plus the emission of a W^- boson. However, the W^- is so heavy that this can only happen by way of Heisenberg's uncertainty principle, for a brief moment in time, when the energy of the W^- is very uncertain. The W^- quickly decays into an electron plus a neutrino. It is this extreme mass of the W^- that makes a weak-interaction process very feeble, relying on a big quantum fluctuation of uncertainty in time and energy. The heaviness of the weak gauge bosons is, in short, why the weak forces are weak (see fig. 36).

Thus there is an enormous difference between the photon and these three new gauge fields. The photon is a massless particle, whereas the W^+, W^-, and Z^0 are each very heavy particles. The forces that are produced by the quantum exchange of W particles between quarks and leptons are exactly the weak forces that Fermi described sixty-five years ago. But what is happening to cause such a discrepancy between the massless photon and the heavy W^+, W^-, and Z^0? How can there be any symmetry between particles of such disparate masses?

ENTER THE HIGGS FIELD

To explain the symmetry breaking of the weak forces, we take our cue from another branch of physics. In the vacuum of free space, the gauge field of electromagnetism—the photon—is perfectly massless. Hence, it always travels at the speed of light. Yet in the laboratory we can make, in a material medium, a kind of "fake vacuum"—something called a *superconductor*. This is a form of *spontaneous symmetry breaking*, like the alignment of magnets, or a pencil falling off of its tip. In a superconductor, which is usually ultracold material such as lead or nickel-niobium, the photon becomes effectively heavy, with a mass of about 1 electron volt. This mass generation for the photon gives rise to the peculiar fea-

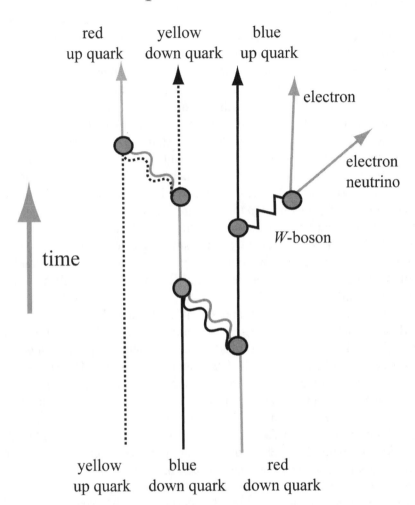

proton

red
up quark

yellow
down quark

blue
up quark

electron

electron
neutrino

W-boson

time

yellow
up quark

blue
down quark

red
down quark

neutron

Figure 36. At the level of quarks and leptons, we see that the process of neutron decay, $n^0 \rightarrow p^+ + e^- + \overline{\nu}^0$ involves the quark transition $d \rightarrow u + e^- + \overline{\nu}^0$ through the exchange of a W gauge boson. The W is so heavy that it cannot be produced with an energy equal to its enormous mass, so it is created for only a brief moment of time with a small energy, allowed by the Heisenberg uncertainty principle. This is an improbable quantum fluctuation, making the weak force very feeble. A free neutron has a half-life of about ten minutes.

tures of superconductors. As noted earlier, superconductors have absolutely no electrical resistance to current flow. Superconductors exist because a "quantum soup" can be made within the ultracold metal that interacts with the photon. This quantum soup has electric charge, and the photon feels that charge, so the photon becomes slightly heavy.

Inspired by the superconductor, we believe that something must change the vacuum of the whole universe to give the large masses to the weak gauge bosons. This something can be modeled by a new field, the wave function of a new particle, that fills all of space. This is called the *Higgs field*, after Peter Higgs of the University of Edinburgh, one of the early researchers who showed how a mathematically modified form of superconductivity might work to explain why the weak forces are so weak and how the electroweak symmetry is spontaneously broken. The strength of the Higgs field in the vacuum is already measured and can be quoted as an energy scale, since it is the theoretical origin of the *Fermi scale*, 175 GeV. At this stage we are postulating a new particle, the Higgs boson, to explain a phenomenon, though we have no real understanding of what it is and where it comes from. We can nonetheless glimpse how the Higgs mechanism works.

All the matter particles and the W^+, W^-, and Z^0, in the Standard Model, would get their masses by interacting with the vacuum-filling Higgs fields (unlike a superconductor, however, the photon does not interact with this particular field and remains massless). The Higgs field is "felt" by the various particles through their *coupling strengths*. For example, the electron has a coupling strength with the Higgs field, g_e. Therefore, the electron mass is determined to be $m_e = g_e \times 175$ GeV. Since we know $m_e = 0.0005$ GeV, we see that $g_e = 0.0005/175 = 0.0000029$. This is an extremely feeble coupling strength, so the electron is a very low-mass particle. The top quark, which has a mass of $m_{top} = 175$ GeV, has a coupling strength almost identically equal to 1, suggesting that the top quark is playing a special role in the symmetry breaking. Other particles, such as neutrinos, have nearly zero masses and therefore nearly zero coupling strengths.

Although this all sounds like a spectacular success, there is a slight problem—there is, at present, *no theory* for the origin of the coupling constants, such as g_e. These appear only as input parameters in the Standard Model. We learn almost nothing about the electron mass, swapping the known experimental value, 0.511 MeV, for the new number, $g_e = 0.0000029$.

The Standard Model successfully and precisely predicted the coupling strength of the W^+, W^-, and Z^0 particles to the Higgs field. These coupling strengths are determined from the known value of the electric charge and another quantity, called the weak mixing angle, measured in neutrino-scattering experiments. So the masses M_W and M_Z (note that W^+ and W^- are just particle and antiparticle of each other and must have the same identical masses; the Z^0 is its own antiparticle) are predicted (correctly) by the theory. The W^+ and W^- have a mass of about 80 GeV, and the Z^0 has a mass of about 90 GeV. These have been measured to very high precision in experiments at CERN, SLAC, and Fermilab.

Symmetry and its spontaneous breaking through the Higgs particle therefore completely controls the mass generation of all the particles in the universe! But hold on—what is the Higgs particle?[10]

This is among the most important scientific questions of our time. The US government, in its wisdom, decided in the 1980s to build a particle accelerator twenty times more powerful than the Fermilab Tevatron in order to thoroughly explore the Higgs boson or, more generally, discover whatever causes the spontaneous symmetry breaking mechanism of the electroweak interactions and the origin of mass. Unfortunately, for a complex of reasons having little to do with science, the project was cancelled in 1993. Therefore we don't know what the Higgs particle is, at present, and we anxiously await the first hints of it, or something like it, from any experiment anywhere.

Hints of the Higgs could possibly emerge, perhaps from Fermilab's Tevatron (now operating well, after a challenging start riddled with technical problems) or from the Large Hadron Collider (LHC) under construction in Geneva, Switzerland, at CERN (the large high-energy physics laboratory of Europe). The LHC is scheduled to begin operations in 2008, colliding protons on protons, each having 7 TeV of energy. This machine will essentially be a microscope with seven times more magnifying power than the Tevatron. We recall that Galileo's new telescope had twenty times the power of the unaided eye and made a plethora of revolutionary discoveries. We expect no less a revolution to come from the CERN LHC, although those of us who work at the Tevatron are trying to find the first clues. When the LHC begins operation, the world's most powerful microscope will have moved out of North America for the first time in almost a century.

So, even though we don't know what the Higgs particle is from experiment, many theories nonetheless abound concerning its true identity.

BEYOND THE HIGGS BOSON: SUPERSYMMETRY?

Throughout these pages we have focused on the profound consequences of symmetry, Noether's theorem, and gauge invariance as the grand unifiers of our understanding of the forces of nature. We have stayed mostly with what is known and will leave more elaborate, and speculative, treatments of what may lie ahead to others. However, the discovery of the Higgs boson is of enormous importance to science because of what it is likely to bring with it. In regard to the Higgs boson, some more words are in order.

The Standard Model has now stood for thirty years as a successful description of all known phenomena. That is a pretty spectacular run for a theory of nature, nowadays. To the Standard Model we must append the weakest of all forces, gravity; yet gravity in the particle physics lab plays almost no role. Gravity is conspicuous in the cosmos and in the apple orchard, but it has for many years defied detailed analysis at short distances, where the other forces and their symmetries emerge. Gravity is part of nature, though, and it does contain a geometrical gauge symmetry, so it must fit into a larger picture of everything.

All physicists today believe that there must be a larger enveloping structure of the Standard Model that is associated with the Higgs boson and eventually gravity. In regard to the Higgs boson, it plainly makes no sense that nature would have provided one and only one particle just to make all other particles acquire mass. However, at the present moment, all data is consistent with the hypothesis that there exists only a single undiscovered Higgs particle; the rest of the matter we have already seen. What this new enveloping structure of the Standard Model actually is has not yet been resolved through experiment, the ultimate arbiter of good taste in physics.

On the cosmological side, there is evidence that new forms of matter, lying outside the Standard Model, inhabit our universe—the so-called *dark matter*. This evidence has been around for a long time, based upon gravitational hints of dark matter in galaxies and clusters dating back to the 1950s. There is also more recent evidence of our universe being an *accelerating universe*, driven by pure vacuum energy, similar to but much less than what was required in the theory of inflation. Moreover, the whole question of vacuum energy is a very difficult one, because ever since the early days of quantum mechanics, physicists have gotten the wrong answer when attempting to compute it—wrong by about 120 powers of ten!

Indeed, these are all things about which our Standard Model, at present, is mute. And, on a rainy Sunday afternoon, we could actually make a long list of open questions, each one of which we could call the "most important question of all of science," because we don't yet have their answers. Whereas the search for the Higgs boson is a well-defined research program, many of these other questions are ill defined. How we answer them, or even attack them, will depend upon what we find with the Higgs. We can only be certain that there is much more to come from the observational science of particle accelerators, as well as deep space telescopes like the Hubble Space Telescope.

Speculations as to what may lie beyond the Higgs boson abound. By far, the most prevalent of these (measured in terms of the tens of thousands of scientific papers on the subject) is the idea of *supersymmetry* (abbreviated SUSY). There are compelling reasons underlying this idea, which may ultimately lead to the unification of all the forces in nature at very high energies (short distances), including gravity. There are also some reasons why SUSY is expected to be associated with the energy scale of the Higgs boson, Fermi's scale of 175 GeV, the experimental elucidation of which may soon be at hand. Supersymmetry can naturally accommodate a Higgs boson and offer a partial explanation as to why it is inhabiting an energy scale of a few hundred GeV. There is indirect evidence that SUSY is more consistent with the idea of a "grand unification" of all forces than other approaches.

Supersymmetry is actually a hypothetical extension of our understanding of space and time. It includes additional dimensions of space that are "fermionic," and the dimensions themselves actually behave like spin-1/2 particles (recall that spin-1/2 particles are called fermions). This means that the new dimensions themselves have weird properties. (For example, the photon, which is a boson with a spin 1, when pushed in the direction of a new fermionic dimension becomes a fermion, called the "photino," with a spin of 1/2. Or, a quark, which is a spin-1/2 fermion, when pushed in the direction of the fermionic dimension, becomes a boson, called a "squark," with a spin of 0.) So, supersymmetry predicts that for every fundamental observed fermion (boson) in nature there must exist a corresponding "superpartner" boson (fermion). We don't yet see these "superpartners" in nature, so if supersymmetry is a valid symmetry, something must be hiding it at the relatively low energies in which we make our observations—the "low" energies of every particle accelerator built to date. Supersymmetry is therefore a broken symmetry.

If SUSY is eventually observed, perhaps at the CERN LHC, our lists of all of the elementary particles will be doubled in size—and there will be full employment for particle physicists. Every particle will have a so-called superpartner. Supersymmetry offers tantalizing possible "dark matter" particles, which can explain the cosmic observations of large quantities of unlit matter within galaxies that is not incorporated into stars and is not shining (hence the rubric "dark matter"). If it exists, SUSY would lend compelling weight in favor to the idea of *superstring theory*, as the ultimate grand unification of all forces, including gravitation.

Superstring theory is the best candidate for a theory of everything. We recall our example of the vibrating guitar string, a metaphor for a quantum motion of an electron in a one-dimensional potential ditch. Superstring theory is the hypothesis that the electron, and all the other elementary particles as well, are literally seen to be vibrating strings. To see this stringy structure of matter would require a microscope that is one hundred thousand trillion times more powerful than the Tevatron, the current most powerful particle accelerator.

What persuades some theorists to believe in a string structure of all matter? It is that string theory solves the problem of incorporating gravity into the overall quantum fabric of nature. The lowest mode of vibration of the quantum string is the graviton. All matter couples universally to gravity, and string theory leads off with this fact. And all of the gauge symmetries and the forces of nature can fit into this picture.

String theory requires supersymmetry. It does not necessarily require SUSY within reach of the Tevatron or LHC, however. Should SUSY be discovered in the laboratory, on the other hand, the strongest possible vote of confidence will have been cast for superstring theory.

There are an infinite number of possible supersymmetric models of the weak scale, that is, at the scale of the Higgs boson mass, but one has become a standard: the minimal supersymmetric Standard Model (MSSM). The MSSM predicts that all of the superpartners of the quarks, leptons, and gauge bosons should be observable, and soon. The MSSM also predicts five observable Higgs bosons. The MSSM is fairly specific about where the lightest Higgs boson must be in mass, and it places it in a well-defined mass range of less than about 140 GeV, accessible to Fermilab's Tevatron with enough collisions, and certainly its successor, the LHC.

The only problem with SUSY is that it offers no real explanation for much of the pattern we see in the masses and properties of the known matter particles (with the possible exception of the heavy top quark).

These properties are relegated to string theory, and the unknown way in which many different symmetries must be broken at inaccessibly high energy scales—this is a physics that only the human brain can access, since it is simply too high an energy scale to ever be witnessed by a particle accelerator in anyone's future.

If SUSY is not seen at the energy scale of the Higgs particles, then it does not imply a death blow to the general idea, since SUSY, as a theoretical construct, can and will retreat to higher and less accessible energy scales, where it cannot be so easily detected. And, even if SUSY has nothing to do with the real world, it has taught us so much about the mathematics of quantum mechanics that it will remain a viable tool for thought for the indefinite future. There is a great deal of intellectual capital riding on the discoveries of the coming decade.

If SUSY is not discovered at the weak scale, then the Higgs boson is likely to be a dynamical entity, perhaps associated with new forces in nature. For example, many theorists have studied the possibility that the top quark, together with such additional new forces, may play a more intimate role in establishing the scale of the weak interactions and thus the masses of all the elementary particles. Here the Higgs particle may be a bound state, perhaps containing the top and anti–top quarks held together by new gauge interactions. If true, such a dynamical scheme would redirect our thinking in completely novel directions. Again, experiment will be the ultimate arbiter.

PHILOSOPHICAL COMMENTS

High-energy physics, the study of the forces and the behavior and structure of matter at the shortest distances, is the ultimate microscopy. Its laws rule the entire universe. In a very real sense, we are now examining and coming to understand the very "genetic code" or the "DNA" of matter itself. What could be more fundamental? The question of the breaking of electroweak symmetry and the origin of mass may be resolved by experiments that are not far away, within the next decade, and probably requiring the energy scale of the Large Hadron Collider at CERN. And, beyond the LHC, a very large hadron collider may be built someday, perhaps in the Gobi Desert of China.

We expect that a major revolution is coming to the science of elementary-particle physics. In the past, such revolutions have also contributed to the knowledge and to the betterment of the human condition around the globe. Throughout the twentieth century, the United States in particular has enjoyed the bountiful harvests derived from physics, chemistry, and biology, due to its great universities and leading laboratories in all scientific disciplines. The impact of the future discoveries at the energy frontier are impossible to predict—that is the nature of fundamental research. Yet we have no reason to believe that we have reached a point of diminishing returns on investments made in fundamental research. In any case, the coming decade will be an exciting one for the human quest to understand the deepest secrets of the universe.

Richard Wagner's Ring Cycle ends with *Götterdämmerung*—the "Twilight of the Gods." Brunhilde leaps to her death on her trusted steed Grane, restoring the golden ring to the Rhine, the source of so much trouble throughout the previous fifteen hours of the opera cycle. In this stirring immolation sequence, Valhalla burns, and the gods are annihilated, their legacy to survive only in the tortured fables of the mortal humans. The humans survive, albeit suffering and not quite comprehending, but going on in their earthly pursuits.

Perhaps this is a metaphor for "separation pain," the urge to reconnect with our protectors, or an admonishment to behave one's self lest we risk perishing. Or perhaps it reflects an intellectual imperative, to eat the apple and get out of Eden for a reason. But we're destined to get beyond the mere fables of creation and to replace them with something else—something more enduring and more rational.

When James Levine performed the Ring Cycle at the Metropolitan Opera over a decade ago, the destruction of Valhalla was depicted as a supernova. Its cataclysmic explosion illuminated the nighttime sky, then faded into a solemn star field, to the bewilderment of the mortals below. It was an awesome visual and musical moment.

When we began the story of symmetry, we noted that the "Twilight of the Gods"—the end of the "Titans"—is also a metaphor for nature's astrophysical behemoths: one hundred solar-mass stars, the Titans of the galaxies, burning brightly and rapidly exhausting their fusion fuel, ultimately being destroyed in an immolation sequence second only to the big bang. All of this is brought on by one of nature's weakest forces and its smallest particles, orchestrated by the deep symmetries that define them and their dynamics. What is meaningful and enduring beyond this particular *Götterdämmerung*? What is the lesson in this for the mortal humans?

The eternal laws of physics, reflected in the human intellect, will go on. Similarly, the quest to understand it all will continue and probably never end, as long as we exist. An all-encompassing "theory of everything" likely doesn't exist (famous last words?)—there will always be an unprovable theorem, or an unknowable higher-energy scale that the accelerator can't reach, a limit to human consciousness, or a translucent veil over a moment of creation through which only shadows are visible. Nature goes on, however, with its eternal laws, permitting us, so far, to see only part of the whole. Although the theory of everything still eludes us, the language has been learned—whatever new answers are found, and deeper questions spawned, about the universe or its mathematical fabric, at the center will be symmetry.

AN EPILOGUE FOR EDUCATORS

T he world we live in today is one of daunting complexity. The challenges are more difficult and urgent than ever. They seem, at times, overwhelming. The means available to solve the world's problems exist, but they involve advanced technologies that are often well beyond the grasp of most people. Therefore we must not only do something, fast, to counter the declining participation and understanding of the technological fields of science and engineering; we must provide a better and more enriched view of the key issues and of what science is—how its natural philosophy works, based upon logic and reflected in the laws of nature. In fact, our future depends critically upon it.

Any visitor to our beloved Fermilab, if not careful, will trip over symmetry upon entering the front door of the lab. The concept of symmetry, known to the ancients, achieved its dominant influence through Albert Einstein's special theory of relativity. In 1905, Einstein recognized the beauty and simplicity of symmetry—which enhances and embellishes architecture, sculpture, and music—as crucial in describing the world.

Now we see symmetry as the centerpiece of a great dining hall's table, at which is seated both classical and modern physics, mathematics, and philosophy, side-by-side, with the beauty and harmony that surrounds us in all forms, in nature, music, and art. Emmy Noether sits at this table, side-by-side with David Hilbert and Einstein. She bequeathed to us one of the most penetrating insights of human knowledge—her remarkable theorem, so important to understanding the dynamical laws of nature. She was one of the greatest mathematicians in history, yet quiet, ascetic, and kind. Few outside of the precincts of mathematics and physics have ever heard of her. Here is a role model for everyone.

Fundamental physics, that of the inner space of particles and the outer space of astrophysics and cosmology, is today in a combined state of utter confusion and breathless excitement. The very air in the relevant laboratories and universities vibrates with expectations of dramatic breakthroughs in the understanding of the history and evolution of the universe and of uncovering the next layer of the defining principles of nature. We are certain that the written laws of physics will be significantly different in ten years compared to what they are now.

This book evolved, originally, out of a program to convince high school science teachers to include some of the important concepts of symmetry in the core disciplines of physics, chemistry, and biology. We initially thought to write a few modules defining symmetry, conveying—at least to the teacher—the reasons that symmetry has come to dominate the frontiers of modern physics and to bring important and fruitful concepts to the physics, chemistry, and biology classroom. Then we set up a Web site, http://www.emmynoether.com, to disseminate the materials more widely. But ultimately this goes well beyond our desire to see a better physics course taught in high school or in the lower levels of a college curriculum. We believe that the education of the general public in science is as noble and essential an ambition as the education of students, and so we enlarged our scope. We intended thereby to serve both ambitions. So many of our young students and senior friends and colleagues who have become interested in all the fuss over superstring theory and modern cosmology can find some place to start. They can now start here.

We believe the ideas that we have conveyed in this book will be essential for the scientific literacy of our students and their teachers—as well as the general public—to appreciate where science is heading. We know from our own experience of the passion all people, young and old, have for stories about antimatter, black holes, neutrinos, and quarks. Now we add to these traditional "hot-button" subjects the variety of symmetries, of space and time, of the gauge symmetries, of supersymmetry, of the CP defiance, and of many other slices of the bread and butter of modern theoretical physics.

By giving our readers some glimpses of the scientists at work—Noether, Einstein, Maxwell, Bohr, Fermi—we hope, via the leitmotif of symmetry, to stress that progress in science depends on the imagination, the inspiration, and the dedication of scientists. It was our motivation to share our enthusiasm and to convey the stories, the sense of adventure and discovery, and above all, the way of thinking that scientific experience—however vicariously—can offer to the general reader.

It is our belief that this way of thinking, with which science both liberates and confines us, should be taught in our schools from kindergarten through high school. For all students, if they are embedded in a seamless set of coherent science and mathematics studies, the scientific way of thinking will emerge to prepare and guide the graduate for all possible futures.

And symmetry, the framework upon which our scientific canvases are stretched, will add the aesthetic, the priceless flashes of clarity and, hopefully, the sense for all readers that, of course, this is the way the world must be.

Ponder what we as a people are trying to do. Through the fog, we will ever try to see how symmetries mold our thoughts and equations and ultimately give form to our conviction that their magic and rhythm—even their imperfections—will reveal, as the fog slowly settles, the beauty and elegance of the universe in which we live.

APPENDIX
Symmetry Groups

THE MATHEMATICS OF SYMMETRY

L et us think concretely about the symmetries of a very simple geometric object—the equilateral triangle. This is a triangle with three equal-length sides, with each side joining another at one of three points called *vertices*. The equilateral triangle presents us with one of the simplest, yet nontrivial, examples of a symmetry. We can draw equilateral triangles on almost any surface, using any colored pens or crayons we wish. We can make the triangles as large or as small as we wish. We can place the triangles anywhere in space or in any orientation in space we wish—for example, with a vertex on top, or with a vertex on the bottom, or whatever.

All such equilateral triangles, irrespective of color, size, position, orientation, or whatever, share a common abstract feature their unique symmetry—which is the defining symmetry of the equilateral triangle, or what it means to be an equilateral triangle. If we could somehow communicate the essence of the symmetry of an equilateral triangle to Martians, they could reconstruct what we are talking about, but they wouldn't know how big or what color or what position of equilateral triangle we are communicating to them. It doesn't matter—the particular symmetry is the essence of what it means to be an equilateral triangle, so let us find a non-visual way to describe it.

It is useful to approach this experimentally (if you can visualize the following manipulations, that's fine—but you are encouraged to actually perform these little experiments yourself or to demonstrate them to a classroom, etc.). We draw two identically sized equilateral triangles (see fig. A1), each on a fairly transparent piece of paper. Alternatively, if possible, draw each figure on a clear plastic overhead projector transparency. These could even be drawn as separate "objects" in a graphical computer editing program that allows the objects to be dragged around, rotated, or overlaid, and so forth.

The triangles are sized so that we can overlay them, matching up the three sides and the three vertices exactly. Let's imagine that the "reference triangle," the one with the three coordinate axes, is taped down so that we can't inadvertently jostle it and change its position. The reference triangle should be considered to be our "coordinate system," and it serves as a "control" in our experiment. Once it is laid in place, we will not move it again. The "experimental triangle," with the ABC labels on its vertices, is our "variable." We can freely move the experimental triangle around, and we can overlay it on top of the reference triangle, matching up the vertices and the sides exactly. We have labeled the experimental triangle's vertices A, B, and C and the reference triangle's axes I, II, and III, just to keep track of what we will do to determine the symmetry operations.

Now we begin our experiment. We first overlay the experimental triangle on the reference triangle, with the vertices reading ABC clockwise, with A at the top, around the experimental triangle. This will be called the *initial position.* We wish to find *all possible distinguishable ways* in which we can pick up the experimental triangle and bring it back down on top of the reference triangle. Each such operation is called a *symmetry operation* or a *transformation.* How do we proceed?

We first rotate the experimental triangle until the vertices read CAB clockwise from the top. This certainly allows us to overlay the experimental triangle on top of the reference triangle and corresponds to a symmetry operation, which is a rotation through 120 degrees (or, equivalently, $2\pi/3$ radians). We'll designate this first symmetry operation as R_{120}. We can therefore start to make a list of symmetry operations, and this one is our first discovery.

It is useful to think of our experimental setup as a kind of "pocket cal-

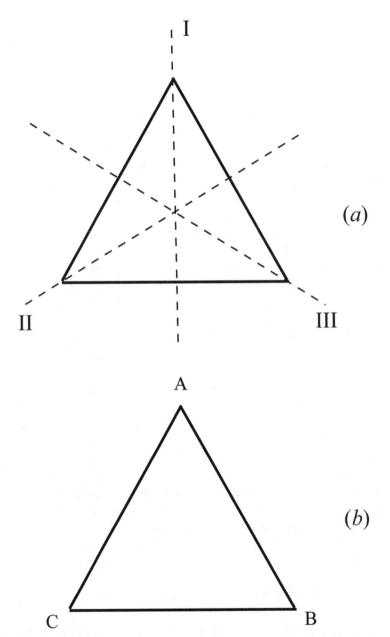

Figure A1. The two equilateral triangles used in our experiment: *a*, the "reference triangle," with axes labeled I, II, and III; *b*, the "experimental triangle," with vertices labeled A, B, and C.

culator." We can *reset* the experimental triangle to the initial position, ABC. Returning to the initial position is like hitting the "Clear" button to start the next calculation. What else might we do to perform a *distinguishable* symmetry operation? It is obvious that a rotation through 240 degrees (or $4\pi/3$ radians) is another symmetry operation, and it yields the result BCA. Thus we discover a second distinguishable symmetry operation, which we designate R_{240}. We can add this one to our list.

Are there other symmetry operations? Perhaps you've considered a rotation by −120 degrees (or $−2\pi/3$ radians), which we might call R_{-120}. We see that this operation takes the triangle to BCA (from the initial position, of course), which was the same result we just obtained for R_{240}. Should we consider this to be a new and distinct symmetry operation, or is R_{-120} equivalent to R_{240}?

In fact, if we were to distinguish between operations like R_{-120} and R_{240}, or R_{-480} and R_{600}, and so on, therefore adding them all to our list, we would really not be focused on the *intrinsic symmetries* of the triangle. Instead, we would be focused more on the *path* we take when we perform the symmetry operation. For example, we could perform an operation such as R_{240} by picking up the experimental triangle and placing it back down, rotated by 240 degrees in the usual way, with the vertices as BCA. Or we could do it by picking up the experimental triangle and then go outside and run around a tree in our backyard ten times, come back in the house, eat a doughnut, and then place the triangle back down with vertices BCA. Is this the same symmetry operation or not? Obviously, there is no added content to the analysis of symmetry of the equilateral triangle by adding in all of this running around in the backyard to the path we take in performing the operation—the triangle symmetry doesn't depend upon whether we run around the tree ten times or seven times, or whether we ate a ham sandwich instead of a doughnut. Indeed, we can barely keep track if we did an R_{240} or an R_{-480} or any R_X, where $X = 240° + 360°N$ and N is any positive or negative integer, after running around the backyard—the path we take is of no interest; only the initial and final positions of the triangle matter.[1]

Therefore we should think of one and only one symmetry operation:

$$R_{240} = R_{-120} = R_{480} = R_X, \text{ where } X = 240° + 360°N,$$

which brings the vertices into the position BCA. This is the essence of the "maximal distinguishability" possible for the symmetry operations. So we will designate this single symmetry operation as R_{240}.

What about a rotation through 360 degrees (or 2π radians)? First, we see that this maps the triangle from the initial position ABC back to the initial position ABC. Indeed, this is therefore a symmetry operation, since it is distinguishable from the others we've considered so far, and it is a very special one at that. For one, it is equivalent to doing *nothing at all*; as such we shall refer to it as the *do-nothing operation*, or the *identity operation*. We shall denote it by the number 1. Second, note that the identity element is a symmetry operation of any object. Even a lowly amoeba, or a pile of rocks, has the identity as a symmetry operation. Finally, note that we could take any path we want, so it is not possible to distinguish between a rotation through 360° or 360° × N for any positive or negative integer N. All are equivalent to the identity operation.

We have now discovered three distinguishable symmetry operations of the equilateral triangle. Are there more? Again, we reset the triangle to the initial position ABC. Consider now a *reflection* about one of the three axes of the reference triangle. We do this by beginning in initial ABC position, and we imagine "skewering" the experimental triangle (as if we had a barbecue skewer and the triangle was a big slab of beefsteak) along one of the axes of symmetry. For example, skewering along axis I, we then pick up the triangle, flip it, and overlay it on the fixed reference triangle. We arrive at the new position, ACB. We denote this symmetry operation as a reflection about axis I, and we'll give it a symbolic name as well: R_I. Similarly, we return to the initial position and consider the other two reflections (1) about axis II, which we call the operation R_{II}, which yields the position BAC, and (b) about axis III, or R_{III}, which yields the position CBA.

At this point we have arrived at the following list of symmetry operations:

1	"do nothing" or "identity"	ABC
R_{120}	rotate by 120°, or $2\pi/3$ radians	CAB
R_{240}	rotate by 240°, or $4\pi/3$ radians	BCA
R_I	reflect about axis I	ACB
R_{II}	reflect about axis II	BAC
R_{III}	reflect about axis III	CBA

Are there any other symmetry operations? At this point we recognize that we have discovered essentially the six permutations of three objects, 3! = 6, i.e., the six permutations of the three vertices of the triangle. Since the vertices must come back down on top of each other under a symmetry operation, each symmetry operation is therefore a permutation of the vertices. There clearly cannot be more than the six operations we have found, so we conclude that we have indeed discovered all possible symmetry operations. This, however, raises an interesting question:

> Q: Are the symmetries of all such objects, such as squares, pentagons, hexagons, cubes, etc., given by the permutations of their vertices?
> A: The answer is *no!*

Whereas all symmetry operations are permutations of vertices, not all permutations of vertices are symmetry operations. We can easily see this in the case of a perfect square. Suppose we have a square with vertices labeled ABCD. A typical valid symmetry operation of the square is a rotation through 90 degrees and gives for the vertices the new position DABC, which is, indeed, also a permutation of the vertices. However, we ask, "Is there a symmetry operation that can give the vertex ordering BACD?" Think in terms of an "experimental square," such as a square sheet of paper, and imagine what we would have to do to get BACD starting from ABCD. We would have to *twist* the experimental square, interchanging vertices A and B but not C and D, to get the vertices into this position, but then the sides of the square would not overlay properly. Since the sides don't overlay, this cannot be a symmetry operation of the entire square. For the square there are only eight symmetry operations, which comes from 4!/3, where we divide the total number of vertex permutations by 3, because there are three classes of twists (do nothing, horizontal twist, and vertical twist). Thus, while all symmetry operations are indeed permutations, not all permutations are symmetry operations. The equilateral triangle is simpler and it does have only six symmetry operations, the ones we've listed above, which are equivalent (*isomorphic*) to the permutations of three objects.

In summary, therefore, with our little bit of experimentation (or play), we have seen that there are only *six different ways* in which we can reposition the top triangle on the bottom triangle. These six different overlay positions of one equilateral triangle on top of another represent the six *symmetry operations* of the equilateral triangle. They are *operations* or *transformations* in the sense that we can start in one overlay position and *operate* on the system by picking up the top triangle and bringing it back into another perfectly overlaying position. In general, we might ask, "How can we be sure that our list of symmetry operations is complete for any given object?" Counting the number of operations can be difficult. Is there another way to do it?

Thus far our experiment has been almost trivial, but now we make a profound observation. We now ask, "Can we obtain additional symmetry operations by *combining together* two of the operations previously obtained?" That is, let us select any two of our six operations, say R_{120} and R_{II}. Let us first perform R_{120} on the experimental triangle and, *without returning to the initial position*, perform the second operation, R_{II}. We see that, if we begin in the initial position, performing R_{120} leads to CAB; then, immediately following with R_{II}, we obtain the position ACB. But ACB is not a new position of the triangle, and we see from our above list that it corresponds to R_I. We have therefore discovered an interesting result: first performing R_{120} and following it by R_{II} yields the result RI. We can write an equation for this:

$$R_{120} \times R_{II} = R_I.$$

Here we have introduced the multiplication symbol, ×, which represents the action of performing the symmetry operations in the order indicated without returning to the initial position after doing the first operation. It is easily seen that the combination of any pair of our symmetry operations by multiplication produces another one. We thus say that our set symmetry operations of elements is *closed* under the operation. Thus the combining of two symmetry operations to compute a new operation is something like the multiplication of numbers. In this sense, the "do-nothing operation" is the true identity, since $1 \times X = X \times 1 = X$.

Mathematicians have a name for this set of six abstract symmetry operations of the equilateral triangle: This is called the *symmetry group* of the equilateral triangle. The symmetry group of the equilateral triangle has the designation S_3.

In general, any symmetry is defined by a collection of symmetry operations that form a symmetry group. The abstract properties of the symmetry group are what attract the immediate attention of mathematicians, who solve many problems in geometry and topology by turning them into *equivalent algebraic problems*. We can now ask if these abstract symmetry operations have any special algebraic properties, like numbers.

We have just seen that a symmetry group forms a self-contained algebraic system. We see this by combining two symmetry operations in sequence. This will always generate a third symmetry operation on our list and thus an element of the group. This becomes a form of "multiplication," and we say that the symmetry group is closed under multiplication.

For this simple set of only six symmetry operations, we can therefore write down the full multiplication table of the equilateral symmetry group (see fig. A2).

	1	$R_{(120)}$	$R_{(240)}$	R_I	R_{II}	R_{III}
1	1	$R_{(120)}$	$R_{(240)}$	R_I	R_{II}	R_{III}
$R_{(120)}$	$R_{(120)}$	$R_{(240)}$	1	R_{III}	R_I	R_{II}
$R_{(240)}$	$R_{(240)}$	1	$R_{(120)}$	R_{II}	R_{III}	R_I
R_I	R_I	R_{II}	R_{III}	1	$R_{(120)}$	$R_{(240)}$
R_{II}	R_{II}	R_{III}	R_I	$R_{(240)}$	1	$R_{(120)}$
R_{III}	R_{III}	R_I	R_{II}	$R_{(120)}$	$R_{(240)}$	1

Figure A2. The multiplication table for the symmetry group of the equilateral triangle.

This table should be read like a highway roadmap. Let's compute $R_{240} \times R_{II}$ from the table. The first operation, R_{240}, is an element in the far left-hand column. The second, R_{II}, is an element in the top row. We then look up the result in the table and find that it is R_{III}. Therefore $R_{240} \times R_{II} = R_{III}$. Since the product of any two of the six elements of the group always gives back another member of the group, we say that the group is closed under multiplication.

One of the remarkable features of a symmetry group is that the multiplication table forms a "magic square." Every one of the six elements of the group (symmetry operation) appears once and only once in every row and column of the table. This is true of *all* symmetry groups.

We further find, to our amazement, that the *commutative law of multiplication*, the one that says $3 \times 4 = 4 \times 3$, *does not necessarily apply* to symmetry groups. That is, we can find two symmetry operations A and B such that $A \times B$ does not equal $B \times A$! We can see this by doing an example. We have already computed the product $R_{240} \times R_{II} = R_{III}$. Now we multiply them in the opposite order, $R_{II} \times R_{240}$, which we see yields R_I. The multiplication does not commute.

Performing symmetry operations in opposite orders generally can yield different results for the combined operations. Some groups do have completely commutative multiplication. In this case we give them a special name: *abelian groups*. The general symmetry group, such as the equilateral triangle group, is noncommutative, or nonabelian.

Noncommutativity is an astonishing fact about symmetries and about ordinary rotations—and therefore about nature itself. Noncommutativity is fairly easy to demonstrate with a book. Take a copy of a book—any book will do—perhaps a copy of *The God Particle*, written by one of the authors of this book (LML). We can perform symmetry operations on the book, just as we might on a sphere. The difference is that the book won't come back identical to its initial position, like the sphere does, but will end up in different positions, so we can see the net result of our rotations (see fig. A3).

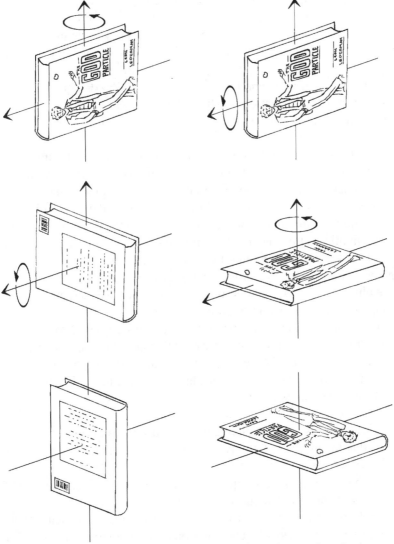

Figure A3. Rotating *The God Particle*. If we perform the rotations in the opposite order, the book ends up in different positions. Rotations in our universe do not commute. (Illustration by Shea Ferrell.)

We can think of an imaginary coordinate system with the book placed at the origin, as shown in figure A3. Now rotate the book through 90 degrees about the imaginary x-axis. We always rotate using the "right-hand rule," i.e., we point our thumb along the positive direction of the axis and rotate in the curling direction of our fingers (the same way a screwdriver is turned to tighten a screw). Call this operation A. Now follow this rotation by another 90-degree rotation about the imaginary y-axis. Call this operation B. Look at where the book ends up; this is the result of A × B. Now go back to the initial position of the book and rotate first along the y-axis (B) and follow it by a rotation along the x-axis (A), and note the position of the book after B × A. Is A × B equal to B × A? The answer is an emphatic *no!* The order in which we perform rotations matters. Noncommutativity is a property of the rotations themselves, not of the object we are rotating.

Our physical world therefore involves abstract forms of numbers that correspond to symmetry operations. These numbers are not like the ordinary numbers, 3, 4, etc. When we multiply, in one order, 3 and 4, we get 12. When we multiply 4 and 3, in the opposite order, we still get 12. Arithmetic is simple in this sense—it doesn't matter what order one uses for the multiplication. Hence, multiplication in arithmetic is commutative. However, the abstract numbers we now encounter can be multiplied together as well, corresponding to successive symmetry operations on some physical system, but their order now matters. We have just seen two examples of noncommutative groups, S_3 and $SU(2)$.

The minimal set of properties that qualifies anything to be a symmetry group has been abstracted by mathematicians who have studied countless symmetries. These properties capture the essence of symmetry and cast it into a set of logical, or algebraic, statements. The properties can be listed as follows.

1. A *group* is a set of elements, X_i, and a composition law, ×, such that the product of any two elements yields another element in the set (closure).
2. Every group has a unique identity element satisfying $1 \times X = X \times 1 = X$ for any element X of the group.
3. Each element of the group has a unique inverse element. That is, given an element X there exists one and only one element X^{-1} such that $X \times X^{-1} = X^{-1} \times X = 1$. (Note that X and X^{-1} can be the same.)
4. Group multiplication is associative: $X \times (Y \times Z) = (X \times Y) \times Z$.

From these statements, or *axioms*, there are many *theorems* that can be proven about groups. For example, the fact that all groups have multiplication tables that form "magic squares" follows from these axioms.

We note that the concept of *associativity* is a bit tricky. It means that, given any three group elements X, Y, and Z, then (in words) we start with the triangle in the initial position and first perform operation Y, and follow it by Z, and remember the result (call this result W), then return to the initial position of the triangle and first do X and follow by W. The result of this sequence of operations will always be the same as having first done X followed by Y then followed by Z. This seems a little complicated, but it is the true operational meaning of associativity.

Indeed, we often take associativity for granted, because the ordinary operations of arithmetic are associative, i.e., $3 \times (4 \times 5) = (3 \times 4) \times 5$. However, there do exist, in pure mathematics, nonassociative systems in which $X \times (Y \times Z)$ does not equal $(X \times Y) \times Z$. These include the case where $\times$ actually represents *division*. That is, when we say "3 divided by 4 divided by 5," we have to carefully specify whether we mean $(3/4)/5 = 0.15$ or $3/(4/5) = 2.75$. So division (when it is not viewed as *multiplying by the inverse* of something) is a nonassociative. As an aside, there are more esoteric things called *normed division algebras*, based upon this idea. These lead to weird new kinds of numbers called *octonions*. Some theorists have attempted to relate nonassociative mathematics to physics. Octonions were thought to be possibly associated with the physics of quarks in the mid-1970s, but the idea never really went anywhere. Nonassociativity seems not to be very relevant to our description of nature. Thus, as far as we can tell, nature is always associative. Symmetries are relevant to nature, and symmetry groups are *always* associative.

Given the defining axioms of groups as the definition of all symmetries, all of the possible symmetries that can possibly exist can be classified by mathematicians. The complete classification of discrete symmetries was, for a long time, a formidable problem, and has only in the past few decades been completed.[2] There are daunting discrete symmetries that exist only in the abstract world. For example, there are crystal lattices that we can imagine to exist in any dimensionality of space that are defined by the *closest packing of spheres* in that dimensionality. One can think of an infinite box full of ball bearings. If we shake the box enough, the ball bearings will fall into a regular crystal lattice. We can ask how many different lattices the ball bearings can form in three dimensions, or four dimensions, or in any dimensionality. These *lattice groups* are discrete symmetry groups with a very large number of symmetry operations.

There are many similarities between the lattices that exist in different dimensions and their symmetries. Remarkably, in exactly twenty-six dimensions, there is a special kind of lattice that occurs in no lesser dimension. It is an *exceptional symmetry*, and it is called the *Monster Group*. There are 8×10^{56} symmetry operations in the monster group.

Finding exceptional symmetries like the Monster Group is the headache that one encounters in attempting to classify all possible discrete symmetries. Solving the problem of finding all possible discrete symmetries involves using computers to prove the enormously complicated theorem of the classification of the discrete symmetries. It is said that no single human being can comprehend the entire proof. Indeed, this is somewhat unsettling, and nowadays there is an entire branch of mathematics that uses computers to try to prove complicated theorems. The novel aspect of proving mathematical theorems by using computers makes a lot of people very uneasy for many various reasons. How does one ever know, even in principle, if the computer got it right? And could it be that the computer will ultimately understand the abstract world better than we humans do? At that point, are we not evolutionarily terminal?

Fortunately, the exercise we just did for the equilateral triangle can be done for any simple geometrical figure. You may want to try it for the square. How many symmetry operations do you count for the square? (Answer: eight, as we saw above.) List them and work out their multiplication table. Then try it for the cube (three-dimensional generalization of a square) or a hypercube (the generalization of a cube into any number of space dimensions). Each of these has its own collection of symmetry operations, which corresponds to its own symmetry group.

A Simple Problem on Sherman's SAT

Let us now ask how symmetry plays a role in a physics problem. Suppose our friendly neighborhood high school student, Sherman, has encountered a physics problem on his practice SAT. Sherman, of course, wants to get a very high grade on the test so that he can enter an expensive university and eventually study to become a lawyer. He is nervous about physics and math questions. Remarkably, however, to understand this physics problem and follow Sherman's steps, we don't actually have to know much about physics or math. We don't have to use any equations, and yet we'll get an idea as to how physical systems actually work in detail and how they are governed by symmetry.

We are given three masses in a triangular formation, with a fourth mass at the center of the triangle (see fig. A4). What is the gravitational force on the center mass exerted by the three other masses arrayed in a triangle?

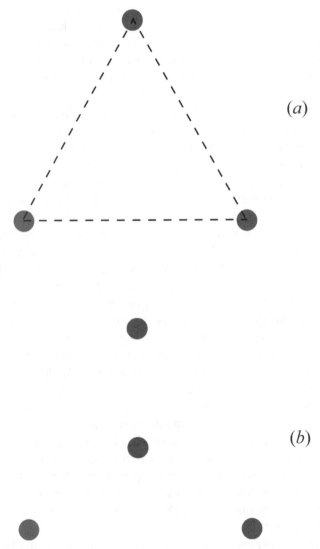

(a)

(b)

Figure A4. Sherman's practice SAT problem: *a*, three symmetrically arranged masses, *b*, a fourth mass at the center. What is the force exerted on the mass at the center? (Diagram by CTH.)

In *a* of figure A4, we have three massive objects arranged in the shape of an equilateral triangle—which should by now be familiar. This means that at each vertex of the imaginary triangle we have placed one of the objects. These objects should each be considered to be at rest. Think of these objects as frozen, or glued, in place. The objects can be almost anything we want, such as billiard balls, planets, black holes, atoms, or very heavy quarks. It is only important that the objects be approximately perfect spheres, or perfect pinpoints of matter, with no internal or hidden structure that we can't see that would mess up the triangular symmetry of our setup. For example, we don't want these objects to be magnets with unseen north and south poles pointing in random directions (it turns out that heavy quarks, like the top quark, have very little magnetism, for a reason associated with something called *heavy-quark symmetry*). We also don't want the objects to be moving, perhaps only instantaneously glimpsed at the very instant they are all aligned at the vertices of the triangle. So, for the particular system of three masses that we are considering, the ordinary equilateral-triangle symmetries that we discussed earlier are exactly the same symmetries of this physical system. We say that the system possesses the symmetry of an equilateral triangle, S_3.

Now, in *b* of figure A4, we have placed a fourth object at the center of the triangular configuration of the objects. Again, it can be any object we wish, but it must have no relevant internal properties that somehow void the symmetry. Again, with the fourth object at the center of the system, we have an exact equilateral-triangle symmetry.

Here's the physics problem that Sherman has to solve: *What is the gravitational force that the center object feels from the three objects at the vertices of the triangle?* Give this a moment's thought and try to answer it for yourself. Remember that a force has both a magnitude, or strength, and a direction in space. Something that has both magnitude and direction is called a *vector*. We usually denote a vector by a little arrow showing the direction it points in space, with the length of the arrow denoting the magnitude of the vector. However, the present problem is pretty simple. Note that if the magnitude of a vector is zero, then the vector itself is zero. If you are guessing the answer to Sherman's homework problem is zero, then you got it right!

Sherman, however, attempts to solve this problem using brute force calculation. Sherman employs the mathematics of vectors and attempts to add together the *force vectors* that each mass exerts upon the center mass. This is a perfectly valid way to calculate the net force on the center object.

Unfortunately, this is a lot of arithmetic, and Sherman gets the result shown in *a* of figure A5.

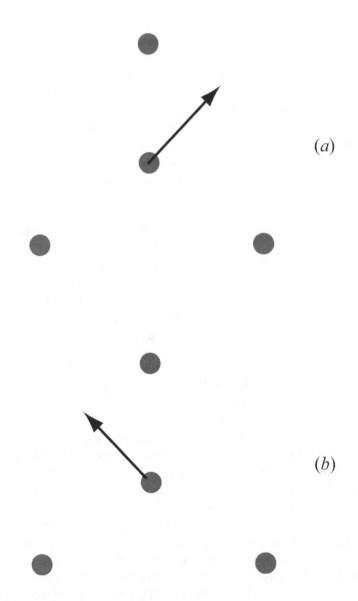

(a)

(b)

Figure A5. Sherman's incorrect result: *a*, Sherman's calculated result for the force vector; *b*, the wrong result when reflected about axis I, a different result. It can't be correct, since it is the same physical system.

There is an easy way to check the solution to this problem using symmetry. Consider Sherman's solution as in *a* of figure A5. Does this solution respect the symmetry of the situation? Remember, if we flip the triangle about symmetry axis I, we must get back the same physical system. However, if we flip Sherman's answer about axis I, we get the result shown in *b*. This is a different answer for the force experienced by the mass at the center. But, it must, by symmetry, be the same identical physics problem in all regards, and therefore it had better be the *same answer!* Therefore Sherman's answer must be wrong.

Sherman, on hearing the news that his answer is wrong, quickly checks his work and finds that he made a mistake. He had lost track of a minus sign in one of his equations and had added two *x*-components of his vectors when he should have subtracted them. Hence, fixing it up, he gets the result shown in *a* of figure A6. The force vector now lies on symmetry axis I. Therefore, if we flip the triangle about axis I, we get back the same answer. It passes the check we did, so maybe this answer is right.

However, the symmetry of the equilateral triangle is a bigger set of transformations than just flips about axis I. For example, we can rotate the triangle by 120 degrees, holding the center fixed in space. Again, if we rotate Sherman's answer we see that it changes, giving the new result, as shown in *b* of figure A6. Therefore this cannot be the right answer, either, because the same physical system must produce the same physics before and after we perform a symmetry operation on it! Sherman, upon hearing this, goes back to his desk and repeats the calculation. He finds yet another error, which he corrects, and, voila! He finally gets the correct answer: zero!

The answer for the force exerted on the mass at the center must be zero based upon the symmetry of the equilateral triangle. Indeed, you may have guessed this at the beginning. The symmetry of the equilateral triangle governs the physics of this simple example and enforces the results in a fundamental way.

But here is an amazing fact: suppose the force we have in mind is no longer gravity, but something else—anything else. For example, the force may be the electric attraction between electric charges. Or it may be the strong nuclear force between protons and neutrons that holds them together inside the nucleus of an atom. Or it may be the force that binds quarks together to make up the particles that compose the nuclei. It may be the gravitational forces experienced by four black holes in a triangular config-

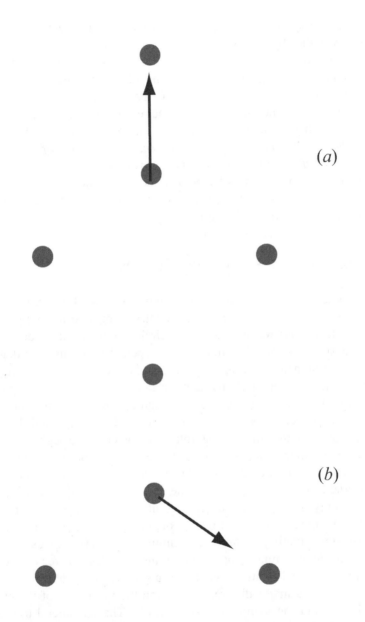

Figure A6. Sherman's correction: *a*, his value for the force vector after correcting his answer so it reflects properly about axis I; in *b*, however, the answer changes under a rotation by 120 degrees. So the answer must be wrong—the only possible correct answer for the force vector is $F = 0$!

uration in space. As long as the equilateral-triangle symmetry is the symmetry of the physical system in our example, then the force at the center is zero, no matter what is producing the force. The symmetry is what makes this happen, not the specific details of gravity, electromagnetism, or quantum chromodynamics. The three particles at the vertex can be black holes, or magnetic monopoles, yet the result remains the same—zero!

Sherman's second SAT practice exam problem is the following: *What is the gravitational force on an object at the center of a hollowed out planet that has the exact symmetry of a sphere?* Sherman didn't even attempt to compute this by brute force; it was now obvious to him that symmetry, again, determines what the force must be. Sherman had learned his lesson well: symmetry is king!

Continuous Symmetry Groups

A circle displays a great deal of symmetry. Indeed, a circle, as we shall see, has an infinite amount of symmetry, compared to an equilateral triangle! Suppose we draw a pair of circles of equal diameter on a pair of transparencies, or fairly transparent paper. We can imagine that our circles are pinned down on top of each other at their centers. We can then reposition the top circle by gently rotating it, with its center pinned. We can rotate through any number of degrees. There are an infinite number of rotations corresponding to any number between 0 and 360 degrees. Most intriguing is that we can rotate the top circle through a tiny, or *infinitesimal*, amount and still have it overlay the bottom circle. There are therefore an infinite number of ways in which one circle can be exactly overlaid on top of another. In the case of the equilateral triangle there were only six such ways, and to reposition the triangle, we had to pick the top one up and move it through a definite step—for example, flipping or discretely rotating it through a minimum of 120 degrees. There is no infinitesimal symmetry operation for the triangles, whereas, for the circle, an infinitesimal rotational change is a symmetry operation.

This is a major difference between the symmetry of the equilateral triangle and the symmetry of the circle. The equilateral-triangle symmetry had no infinitesimal symmetry operations. We had to pick up one triangle and rotate it through at least 120 degrees, or flip it, to get it to overlay the other triangle. With the circle, any rotation will do—even by 0.000000001 degrees of rotation—is a symmetry operation of the circle.

The set of all symmetry operations of a circle also forms a group. Mathematicians have given the symmetry group of the circle a special name too: it is called $U(1)$. We see that this group has an infinite number of operations and no smallest nonzero symmetry operations. On the other hand, we say the symmetry group of the equilateral triangle is *discrete*, because it involves discrete steps and has no infinitesimal operations. We also say that the symmetry group of the circle is *continuous*, because it has infinitesimal operations and an infinite number of symmetry operations.

Because continuous symmetry groups have an infinite number of symmetry operations, or elements, we cannot write down their multiplication table—the number of elements would effectively be infinity times infinity. How, then, might we analyze continuous symmetry groups? Students of calculus might have guessed that the notion of the *rate of change*, or the *derivative*, which applies to continuous functions, will also apply to continuous symmetries. In fact, this leads to the heart of the method of classifying continuous symmetries. Instead of analyzing the full multiplication table, we need only look at tiny, or infinitesimal, symmetry operations (the derivative of a rotation with respect to the angle). These are called the *generators* of the group. From the generators we can reconstruct all of the symmetry operations.

The *generators* thus form their own special, self-contained, system of mathematics, called a *Lie algebra,* after famous Norwegian mathematician Sophus Lie (pronounced "Lee"), born in 1842, who pioneered this technique.[3] By considering the Lie algebras rather than the infinite sets of transformations, all possible continuous symmetries that can exist have been determined. All possible continuous symmetries were classified by mathematicians, particularly by French mathematician Elie-Joseph Cartan, early in the twentieth century.

The symmetry group $U(1)$ has a trivial Lie algebra, consisting of only one generator, since there is only one direction in which we can rotate a circle. The sphere is the next step up, "a circle in three dimensions." We imagine a sphere floating in space, with its center pinned to a particular point in space (for example, imagine a basketball floating in space with its center located at a fixed point). We imagine all of the rotations that we can do that would overlay the sphere on itself, keeping the center pinned. There are obviously an infinite number of such rotations. The basketball, or the sphere, is not like a circle, forced to lie in a two-dimensional plane. Rather, we can rotate the sphere about any line that passes through its center, and all such rotations keep the sphere on top of itself.

For the sphere there are three distinct generators. These are tiny rotations about an imaginary x-, y-, or z-axis. There are again an infinite number of symmetry operations of the sphere. Clearly there are infinitely many more symmetry operations of the sphere than the circle since we can rotate the sphere in three dimensions. We thus say that the sphere has a larger symmetry than the circle. Mathematicians call the collection of symmetry operations that leave a sphere invariant the symmetry group $SU(2)$. (It is equivalent to another group, called $SO(3)$; actually, the equivalence is not exact, since $SU(2)$ contains $SO(3)$, but this is a subtlety into which we won't enter.)

If the three generators of $SU(2)$ are designated as T_x, T_y, and T_z, any operation (rotation) that is an element of the group $SU(2)$ can be written as $\exp(iT_x\theta_x + iT_y\theta_y + iT_z\theta_z) = 1 + iT_x\theta_x + iT_y\theta_y + iT_z\theta_z + \ldots$, where θ_x, θ_y, and θ_z are the *rotation angles* (or "parameters") of the group element. We have written the approximate form for very small rotation angles, and the exponential is defined by its series expansion.

Again, we encounter the intriguing property of noncommutativity. If we perform one rotation of the sphere, call it A, and follow it by another rotation, call it B, we get a result that we can think of as $A \times B$. We can then perform these rotations in the opposite order, and we get $B \times A$. We find that $A \times B$ generally does not equal $B \times A$, the property called noncommutativity. By considering an infinitesimal rotation about the x-axis followed by an infinitesimal rotation about the y-axis, and then considering the opposite ordering, we find that the generators satisfy the relationships $T_xT_y - T_yT_x = 2iT_z$, $T_yT_z - T_zT_y = 2iT_x$, and $T_zT_x - T_xT_z = 2iT_y$. These relations define the Lie algebra of the group $SU(2)$ much like the multiplication table defined the group S_3. This is the analogue of the group multiplication table that existed for discrete groups with a finite number of elements—we now have a finite number of generators that satisfy the noncommutative Lie algebra. The classification problem of all continuous symmetries has now reduced the problem of classifying all Lie algebras.

Combinations like $T_yT_y - T_yT_x$ are so important in group theory and quantum mechanics (even classical mechanics) that they have a special name. They are called *commutators* and are written as $T_xT_y - T_yT_x = [T_x,T_y]$. Multiplication of the group generators is associative.

Quite often a Lie algebra breaks into two or more parts, each part completely commuting with the other part(s). This then represents two or more distinct symmetries. Lie algebras that do not decompose in this way are called *simple groups*. The more complex groups are obtained by "multiplying" two symmetry groups together, in the sense of a Cartesian product. An example of this is afforded by something like a heavy quark, such as the beauty quark, or b-quark. This quark has spin, so it represents the symmetry of rotations $SU(2)$, but it also has quark color, so it simultaneously represents the quark color symmetry of $SU(3)$. Therefore the combined symmetry that applies to b-quarks is the composite, or *product group*, $SU(2) \times SU(3)$, and all of the $SU(3)$ color rotations commute with the $SU(2)$ spin rotations.

The full classification of all possible simple Lie algebras was completed in the early twentieth century and is known as the *Cartan classification*. They are as follows:

1. Rotational symmetries of spheres that live in N real coordinate dimensions: $O(2) = U(1)$, $SO(3) = SU(2)$, $SO(4)$, $SO(5)$, ..., $SO(N)$, ...
2. Rotational symmetries of spheres that live in N complex coordinate dimensions: $U(1)$, $SU(2)$, $SU(3)$, $SU(4)$, ..., $SU(N)$, ...
3. Symplectic groups are the symmetries of N harmonic oscillators: $Sp(2)$, $Sp(4)$, ..., $Sp(2N)$, ...
4. The exceptional groups: G_2, F_4, E_6, E_7, and E_8.

For example, the groups $SO(N)$ are called the "special orthogonal groups" and are the symmetries of spheres that live in N dimensions of space, where the space has real coordinates. They are therefore the set of transformations that map $(x_1, x_2, \ldots x_N) \to (x_1', x_2', \ldots x_N')$, leaving invariant the N-dimensional sphere, $1 = x_1^2 + x_2^2 + \ldots + x_N^2 = x_1'^2 + x_2'^2 + \ldots x_N'^2$. Note that there are reflections that are discrete transformations that leave also the sphere invariant, such as $(x_1, x_2, \ldots x_N) \to -(x_1, x_2, \ldots x_N)$. These are omitted in the definition of $SO(N)$, and that is why we have the S in $SO(N)$ for "special" (technically, the "orthogonal groups," $O(N)$, include these discrete reflections).

On the other hand, the groups $SU(N)$ are called the "special unitary groups" and are the symmetries of spheres that live in N dimensions of space, where the space has complex coordinates. They are therefore the set of transformations that map $(z_1, z_2, \ldots z_N) \to (z_1', z_2', \ldots z_N')$ and leave invariant the equation of the complex N-dimensional sphere, $1 = |z_1|^2 + |z_2|^2 + \ldots + |z_N|^2 = |z_1'|^2 + |z_2'|^2 + \ldots + |z_N'|^2$.

Such symmetries are relevant to quantum mechanics, because the physical state of a system is viewed as a vector in a complex space (this is actually a more fundamental description than the wave function). For example, quark color is a vector in a three-dimensional complex coordinate space in which we call the axes "red," "blue," and "yellow," and the symmetry group of color is $SU(3)$. Here the overall factors of $U(1)$ that are part of the symmetry that are omitted in the definition of $SU(N)$ (technically, the "unitary groups," $U(N)$, include the additional factor of $U(1)$). The symplectic groups have a similar invariance, acting on a $2N$-dimensional vector made of noncommuting numbers.

Finally, we arrive at the famous *exceptional groups*, G_2, F_4, E_6, E_7, and E_8. These groups have no simple obvious symmetry interpretations, but they have remarkable properties. These have always been fascinating symmetry groups from the perspective of the grand unification of the all the fundamental forces in nature that are described by local gauge symmetries. This is because the known forces in nature are described by the gauge groups, $U(1) \times SU(2) \times SU(3)$, and they fit naturally (that is, they are *subgroups*, which are smaller groups nested within larger groups) into $SU(5)$, which in turn fits into $SO(10)$. The group $SU(5)$ was the first compelling grand unified theory, proposed in the mid-1970s by Howard Georgi and Sheldon Glashow. This, in turn, fits naturally into a telescoping set of exceptional groups, $SO(10) \supset E_6 \supset E_7 \supset E_8$.

In the 1980s it was shown by John Schwarz and Mike Green that the largest exceptional group, E_8 (actually, the direct product, $E_8 \times E_8$), is intimately connected to string theory, representing one of the few consistently allowed symmetries of a world described by a quantum-mechanical superstring. This, together with the fact that string theory seems naturally to contain a consistent quantum gravity, has led to the enormous interest in superstring theory as the ultimate theory of all nature's fundamental forces.

NOTES

INTRODUCTION

1. Albert Schweitzer, *J. S. Bach*, trans. Ernest Newman (Mineola, NY: Dover, 1966), pp. 99–101, 227.

2. A biography of Bach, together with an interesting history of musical form and analysis of its relationship to symmetry, is given by Prof. Timothy Smith of Northern Arizona University, "Sojourn: The Canons and Fugues of J. S. Bach," http://jan.ucc.nau.edu/~tas3/bach index.html (accessed May 7, 2004), and "Lüneburg (1700–1703)," http://jan.ucc.nau.edu/~tas3/luneburg.html#french (accessed May 7, 2004). Professor Smith has coined the term *back and forth technique* for some of the complex symmetrical patterns found in Bach's music.

3. See Timothy Smith, "Bach: The Baroque and Beyond; The Symmetrical Binary Principle," http://jan.ucc.nau.edu/~tas3/bin.html#note2 (accessed July 15, 2004). "The imitation of dance in the arrangement of tones, in which phrases are joined together like intertwining steps in the formal dance, the symmetrical arrangement of measures, all this has shaped that creation known as the *pièce*, which represents poetry in music. The entire ambition of the French nation is directed toward . . . that subtle *symmetrical* division, which shapes the musical figures in a *pièce*,

like those ornamental shapes of boxwood hedges which comprise a garden in the parterre of the Tuileries." This quote, attributed to a Bach contemporary, historian Abbot Jean-Bernard LeBlanc, published in Manfred F. Bukofzer, *Music in the Baroque Era* (New York: W. W. Norton, 1947), p. 351.

4. William Manchester, *A World Lit Only by Fire: The Medieval Mind and the Renaissance Portrait of an Age* (Boston: Back Bay Books, 1993), p. 230.

5. Will Durant and Ariel Durant, *The Story of Civilization*, vol. 2, *The Life of Greece* (New York: Simon & Schuster, 1966), pp. 636–37.

6. See the excellent account of group theory and Galois' life in Simon Singh, *Fermat's Enigma* (New York: Walker, 1997), pp. 223–26.

7. Readers interested in the basic mathematical aspects of symmetry can turn to the appendix, where we develop the elementary concepts and exhibit some accessible "gee-whiz" results of group theory. This is ideally suited as an introduction to the subject in a high school algebra or physics class—or a rainy Sunday afternoon of mathematical recreation.

CHAPTER 1

1. For a timeless introduction to the big bang theory, see Steven Weinberg, *The First Three Minutes* (New York: Basic Books, 1977).

2. Hesiod, *Theogony*, trans. N. Brown (New York: Liberal Arts Press, 1953), ll. 116–138; the Berkeley Online Medieval and Classical Library provides a translation and analysis at their Web site: http://sunsite.berkeley.edu/OMACL/Hesiod/theogony.html (accessed May 10, 2004). We have omitted in the main text discussion the detailed account of the *Theogony*, which we summarize here.

Gaia gave birth to the Cyclopes (Greek for "orb-eyed"), gigantic one-eyed monsters, named Brontes, Steropes, and Arges, who represented *thunder, lightning*, and the *lightning bolt* (they ultimately ended up on Mount Olympus and became the blacksmiths of the weaponry of Zeus). Gaia's third litter consisted of the Hecatoncheires, named Aegaeon, Cottus, and Gyges, which were gigantic beasts with fifty heads and one hundred arms, each of colossal strength. Ouranos found the Hecatoncheires so detestable, no doubt an affrontery to his paternalistic pride, that he hid them away "in a secret place of Earth," much to the chagrin of their mother Gaia.

Gaia, fuming over the Hecatoncheires affair, persuaded her Titan son Cronus to overthrow Ouranos. Cronus ambushed his father, castrated him with a sickle, and then became the ruler of the Titans. The excised Ouranos's vital parts were cast into the sea, becoming the sea foam, from which arose the goddess of beauty, Aphrodite (Venus). The Hecatoncheires were permitted to reemerge from sequester.

Cronus, however, was a paranoid ruler. To avoid any possible future usurpation of his supremacy by way of his offspring, he routinely made a practice of swallowing his own children. His wife, Rhea, tricked him into swallowing a rock instead of her son Zeus and the Olympians, the future gods of Hellenistic Greece, thus saving them. Cronus, in a primordial act of sibling rivalry, cast Gaia's grotesque children, the Cyclopes and the Hecatoncheires, down into the underworld, Tartarus. Thus the overthrow of Ouranos did not help Gaia's cause. Tartarus was guarded by Gaia's latter offspring, the monstrous Hecatoncheires.

Ultimately, Zeus revolted against his father Cronus and the other Titans, defeating them. All of the Titans were banished to Tartarus. Cronus, however, managed to escape and settled in Italy, where he ruled as the Roman god Saturn. The period of his rule was said to be a golden age on Earth and is honored in the Roman tradition by the Saturnalia feast. The Titans were followed by the rule of Zeus and his generation from Mount Olympus.

3. Arthur Koestler, *The Sleepwalkers: A History of Man's Changing Vision of the Universe* (New York: Macmillan, 1959; London: Arkana, 1989), p. 35.

4. For more information on the formation of the elements, see "From the Big Bang to the End of the Universe: The Mysteries of Deep Space Timeline," PBS Online Web site, http://www.pbs.org/deepspace /timeline/ (accessed May 10, 2004); "Tests of the Big Bang: The Light Elements," NASA WMAP homepage, http://map.gsfc.nasa.gov/m_uni/ uni_101bbtest2.html (accessed May 10, 2004). Accessible scientific articles include P. J. E. Peebles et al., "The Case for the Relativistic Hot Big Bang Cosmology," *Nature* 352 (1991): 769; and P. J. E. Peebles et al., "The Evolution of the Universe," *Scientific American* 271 (1994): 29. Typing *nucleosynthesis* into a search engine such as *Google* will yield numerous good Web sites. *Big-bang nucleosynthesis* deals with the formation of helium, deuterium, and lithium in the very early universe (before the first ten minutes or so), and *stellar nucleosynthesis* deals with the heavy-element formation in stars. A detailed account of the nucle-

osynthesis of elements heavier than iron can be found at "Nucleosyn-theis," http://ultraman.ssl.berkeley.edu/nucleosynthesis.html (accessed May 10, 2004). The classic scientific references for stellar nucleosyn-thesis are R. A. Alpher, H. A. Bethe, and G. Gamow, *Physical Review* 73 (1948): 803; E. M. Burbidge et al., *Reviews of Modern Physics* 29 (1957): 547.

5. See, in particular, "Oklo's Natural Fission Reactors," American Nuclear Society Web site, http://www.ans.org/pi/np/oklo/; "Oklo Fossil Reactors," Curtin University Center for Mass Spectrometry Web site, http://www.curtin.edu.au/curtin/centre/waisrc/OKLO/index.shtml; and "Oklo," *Wikipedia*, http://en.wikipedia.org/wiki/Oklo (all accessed May 10, 2004).

6. In addition to the information provided in sources named in note 5, for a general discussion of limits on time dependence of fundamental parameters in physics, see F. W. Dyson, "Time Variation of Fundamental Constants," in *Aspects of Quantum Theory*, ed. A. Salam and E. P. Wigner (Cambridge: Cambridge University Press, 1972), pp. 213–36.

CHAPTER 2

1. Descriptions of all things in physics or engineering are built out of three basic physical quantities: length, time, and mass. For example, to specify the *quantity of matter* of some specimen, irrespective of the details of its composition, we give the *mass* of the specimen. The "spec-imen" may be a proton, an electron, a virus, the Eiffel Tower, or the planet Jupiter. We don't have to use one kind of mass to describe protons and another to describe electrons.

We generally deal with more complex quantities in physics. For example, the measure of quantity of motion is the *speed* with which an object moves. Speed is a certain distance traveled within a certain amount of time. Therefore, speed is a length interval divided by a time interval, and we say that speed has the engineering dimensions of L/T. Accelera-tion is the *rate of change of velocity per unit time* and thus has engi-neering dimensions L/T^2. When a massive object moves we find the measure of physical motion is the *momentum*, which is mass times velocity, or ML/T. We also have *energy*, which can occur in almost any form, and which has the dimensions of mass times velocity squared, or ML^2/T^2. Power is the *time rate of change of a quantity of energy*, or

ML^2/T^3. From Newton's equations we learn that a *force* produces *a time rate of change of momentum*, hence a force has engineering dimensions of ML/T^2.

In science, we mainly use one of two systems of units of measurement: (1) the centimeter-gram-second, or *cgs*, system; or (2) the meter-kilogram-second, or MKS, system. We mostly use the MKS system in this book, but this is essentially an arbitrary choice. Mass in MKS is measured in kilograms, length in meters, and time in seconds (for both systems). For the present examples, the following conversions are used: 1 meter = 100 centimeters = 3.28 feet = 1.09 yards; 1 pound on Earth is equivalent to 0.45 kilograms mass; 1 kilogram is equivalent to 2.22 pounds on Earth (note that pounds refers to weight, which is a force, ML/T^2, while kilograms refer to mass, M; on the Moon the weight of an object will change, but the mass will remain the same; the *cgs* system is handy because the mass of one cubic centimeter of water at room temperature is 1 gram). One year contains 3.15×10^7 seconds.

2. Note that the Internet search engine Google can perform many simple unit conversions upon simply asking a question. For example, go to http://www.google.com and type in the search window the question "How many square meters per acre?" When you then hit the button labeled "Google Search," out pops the answer: 1 acre = 4,046.85642 square meters. When it cannot answer a question, it will generally refer you to many other sites that can.

3. Note that the English system uses *foot-slug-seconds*, and a *slug* is a unit of mass equal to a pound of force divided by the value of *g*, or 32; needless to say, not many physicists use the quaint old English system today.

4. See Robert L. Park, *Voodoo Science* (Oxford: Oxford University Press, 2000), pp. 3–14, for other contemporary free-energy schemes. A perpetual-motion machine is generally defined to be a machine that runs indefinitely with no consumption or production of energy, whereas a free-energy machine is one that produces excess energy from nowhere. See "The Museum of Unworkable Devices," Donald Simanek's pages, http://www.lhup.edu/~dsimanek/museum/unwork.htm; and "Eric's History of Perpetual Motion and Free Energy Machines," Philadelphia Association for Critical Thinking Web site, http://www.phact.org/e/dennis4 .html (both accessed on May 12, 2004).

5. Mark Twain, *The Tragedy of Pudd'nhead Wilson*, from *Pudd'nhead Wilson's Calendar* in chapter 14, "Tom Stares at Ruin."

6. What actually happens in a complete physical process is slightly more complicated. There must always be other particles involved in the collision to conserve both energy and *momentum*. Nonetheless, the emission or absorption of a photon by an electron is the fundamental defining process of electrodynamics. We revisit this in detail in chapter 11.

7. David Goodstein, *Out of Gas: The End of the Age of Oil* (New York: W. W. Norton, 2004).

8. As mentioned earlier, mass is a measure of the quantity of matter in an object. The 1,000-kilogram automobile will have a *weight* on the surface of Earth of 2,200 pounds, a little more than one ton. The weight is the *force of gravity* that the object feels at the surface of Earth. Beware, therefore, that pounds and tons are *not* units of mass, but kilograms are! The same car on the Moon will still have a mass of 1,000 kilograms but a weight of only 370 pounds. In free fall in space, the car will still have a mass of 1,000 kilograms but a weight of zero. In physics you should always forget about weight and think *only* in terms of mass. See note 1 above.

9. To get the approximate speed in meters per second, divide the velocity in miles per hour by 2, useful for rough estimates. Or, go to http://www.google.com and type "60 miles per hour"; then hit the button labeled "Google Search" and read out "26.8224 m / s," which we are rounding up to 30 meters per second. Throughout this book, we will use "physicists' rough estimates" for things; the most important thing one can do to understand the physical world (or socioeconomic world, for that matter) is to be able to do what are called order-of-magnitude estimates of things.

10. We're using the simple formula for the kinetic energy of an object in Newtonian physics, $E = Mv^2/2$. To compute E we must always use one consistent system of units, and the MKS system is one of the most convenient. See note 1 above.

11. You may want to try this little experiment with a sport utility vehicle and then compare the result to a compact car or a motorcycle. You will need to know the masses of these vehicles. For example, going to http://www.new-cars.com, we found that a certain five-speed compact imported car with manual transmission weighed 2,590 pounds, so, dividing by 2.2, had a mass of 1,177 kilograms (not counting driver and gasoline). Performing the experiment, we found that it took about 10 seconds to slow from 60 mph to 50 mph on the open highway, so the compact car was consuming a power of about 16,100 watts, or 16 kilowatts.

Now, one gallon of gasoline contains about 110,000,000 (110 million) joules of chemical energy. At 60 miles per hour, or 1/60 of a mile per second, our compact car is consuming 16,000 watts/(1/60 miles per second) = about 1 million joules per mile. If the engine were 100 percent efficient, we would be getting a gas mileage of (110 million joules/gallon)/(1 million joules per mile), or about 110 miles per gallon. However, our car's gas mileage indicator read 35 miles per gallon. Therefore we find that the automobile had an efficiency of about 32 percent. Trying this experiment yourself with an SUV, or even a motorcycle, may illustrate the difference in gasoline consumption of these vehicles and why you or your neighbor, driving the SUV, is paying a fortune every time one of you stops at the gasoline pump.

12. This is often quoted in BTUs, which is another unit of energy. The annual energy consumption of the United States, at the present time, is about 100 quadrillion (10^{15}) BTUs, or a total of 10^{17} BTUs (see, for example, the US Department of Energy Web site, http://www.eia.doe .gov/emeu/aer/diagram1.html [accessed July 16, 2004]). One BTU is approximately 1,000 joules; hence, a total of 10^{20} joules are consumed in the United States per year. There are about 300 million (3×10^8) US citizens and about 30 million (3×10^7) seconds in a year, so the average American is consuming about $10^{20}/(3 \times 10^8 \times 3 \times 10^7)$ watts, which is approximately 10,000 watts. Personal consumption, in a family's home or daily activity, we take to be roughly 3,000 watts. This is about five times greater than the world average.

13. Many people believe the answer to our energy needs lies in going back to basics, such as using wood-burning stoves. There are serious efforts under way to develop biomass fuels in a science known as "bioenergy." Vast tracts of land can be planted in graceful and slender poplar trees, cottonwoods, and willows, which are effectively a form of solar collector. Putting some numbers together, one gets a rough estimate of a power yield from trees of about 1 watt per square meter. That is a solar efficiency of about 1 percent, marginal for a society with high per capita power consumption. In addition, the atmospheric pollution from burning wood is not small. See the Bioenergy Information Network, http://bioenergy.ornl.gov (accessed May 14, 2004), which provides various useful bioenergy measures and conversion factors.

14. For more information, see the International Thermonuclear Experimental Reactor Web site at http://www.iter.org (accessed May 14, 2004).

CHAPTER 3

1. It is, however, an open question whether or not supplemental information is required to specify the conditions at the initial instant, $t = 0$, of the universe. It has been argued that a smooth extrapolation of the laws of physics to this instant may be meaningful and sufficient, in J. B. Hartle and S. W. Hawking, "Wave Function of the Universe," *Physical Review* D28 (1983): 2960.

2. Robert K. Massie, *Dreadnought* (New York: Ballantine Books, 1991), pp. 38–43.

3. Quoted in Simon Singh, *Fermat's Enigma* (New York: Walker, 1997), p. 100.

4. Notice from the 1916–17 University of Göttingen course catalogue, cited in the excellent Emmy Noether biographical article by J. J. O'Connor and E. F. Robertson at "Emmy Amalie Noether," http://www-gap.dcs.st-and.ac.uk/~history/Mathematicians/Noether_Emmy.html and http://www-gap.dcs.st-and.ac.uk/~history/PictDisplay/Noether_Emmy.html (both accessed May 14, 2004). The latter site has an extensive photo gallery. Finding publishable photographs of Noether is made difficult because ownership information is generally unavailable.

5. Nina Byers, "E. Noether's Discovery of the Deep Connection between Symmetries and Conservation Laws," presented at the Symposium on the Heritage of Emmy Noether in Algebra, Geometry, and Physics; published in *Israel Mathematical Conference Proceedings* 12 (1999); see also Nina Byers, ed., "Emmy Noether: 1882–1935," Contributions of 20th-Century Women to Physics, http://www.physics.ucla.edu/~cwp/Phase2/Noether,_Amalie_Emmy@861234567.html (accessed May 14, 2004).

6. Emmy Noether, *Collected Papers*, ed. Nathan Jacobson (New York: Springer Verlag, 1983).

7. In a nutshell, Gödel proved that any mathematical system will always contain "theorems" that cannot be proven to be true or false. During Hilbert's era it was shown that mathematics itself is logically equivalent to arithmetic. The axioms, or starting assumptions, are analogous to a chosen set of prime numbers. For example, a given mathematical system may contain five axioms corresponding to the primes (2, 3, 5, 7, 11). The provable theorems are represented by the numbers that can be factored into this set of primes. For example, the theorem corresponding to the number 44 can be proven because $44 = 2 \times 2 \times 11$, and 2 and 11 are axioms in the system. However, the theorem corresponding to the number

17 cannot be proven, because 17 cannot be factored into the set of axioms we have chosen, the set of five primes up to 11. Any mathematical system containing a finite number of axioms is therefore "incomplete"—the content of Gödel's theorem. This raised the specter that certain of the great theorems on Hilbert's famous to-do list were, in fact, not provable. Until recently, even the defiant Fermat theorem was believed to be a candidate for Gödel's incompleteness, but it was heroically proven in 1993 by Andrew Wiles of Princeton University (see Singh, *Fermat's Enigma*).

8. Hermann Weyl, "Emmy Noether," *Scripta Mathematica* 3 (1935): 201–20.

9. Other interesting biographical sources include: Clark Kimberling, "Emmy Noether (1882–1935): Mathematician," Clark Kimberling home page, faculty.evansville.edu/ck6/bstud/noether.html (accessed on May 14, 2004); Auguste Dick, *Emmy Noether, 1882–1935*, trans. H. I. Blocher (Boston: Birkhauser, 1981); C. Kimberling, "Emmy Noether," *American Mathematical Monthly* 79 (1972): 136–49; Martha K. Smith, *Emmy Noether: A Tribute to Her Life and Work*, ed. James W. Brewer (New York: Marcel Dekker, 1981); C. Kimberling, "Emmy Noether, Greatest Woman Mathematician," *Mathematics Teacher* 75 (1982): 53–57; Lyn M. Olsen, *Women of Mathematics* (Cambridge, MA: MIT Press, 1974), p. 141; Sharon Bertsch McGrayne, *Nobel Prize Women in Science: Their Lives, Struggles, and Momentous Discoveries* (New York: Carol, 1993).

10. Albert Einstein, "The Late Emmy Noether: Professor Einstein Writes in Appreciation of a Fellow Mathematician," *New York Times*, May 4, 1935, p. 12.

CHAPTER 4

1. Here's a partial list of some of the basic parameters of nature. These are a few of the things that Gedankenlab might measure as it drifts throughout our universe:

Speed of light	c	=	2.99792458×10^8 m/s
Planck's constant	$\hbar$	=	$1.054571596 \times 10^{-34}$ m^2 kg/s
Newton's constant	G_N	=	6.673×10^{-11} m^3/kg s^2
Unit of electric charge	e	=	$1.602176462 \times 10^{-19}$ coulombs
Mass of electron	m_e	=	$9.10938188 \times 10^{-31}$ kg
Mass of proton	m_p	=	$1.67262158 \times 10^{-27}$ kg

2. Remarkably, there is one observation that is slightly puzzling in light of these results. Gedankenlab finds that the matter in the universe does seem to have a preferred state of motion, while the laws of physics are independent of one's state of motion. That is, there is a special state of motion, a particular velocity, relative to which the average motion of all the galaxies and the relic thermal radiation in the universe is zero. Of course, any given galaxy is moving with some random velocity, but taken together the galaxies define a special state of motion. However, the fundamental laws of physics, embodied in measurements of the speed of light, the mass of the electrons, and all other physical quantities do not depend upon the lab's state of motion.

3. Often we take the *absolute* (or positive) value of the difference and define the length to be $L = |x_{tip} - x_{handle}|$. Later, when we discuss relativity, we will define the *separation* as $L = x_{tip} - x_{handle}$, which is essentially the length but which can be negative.

4. See C. T. Hill, M. S. Turner, and P. J. Steinhardt, "Can Oscillating Physics Explain an Apparently Periodic Universe?" *Physics Letters* B252 (1990): 343–48 and references therein.

5. There is a question that bothers many people at this point, and it has to do with the nature of how physics describes things. We have seen that both space and time translational symmetries are valid. Yet, although I can translate myself through space, I seem to have no freedom to translate myself through time. Time travel belongs to science fiction, not to reality. Moreover, although I can view a mountain range and view the three dimensions of space, I only experience one instant of time—I cannot view all of history like a mountain range (like Kurt Vonnegut's Trafalmadorians), nor can I freely traverse this temporal mountain range, as I can a spatial one. I can recall events of the past but cannot recall events in the future. I am, like the Chinese say, "backing into the future," since my "eyes" can only see the past. Why is the *perception* of time so different than that of space? The keyword here is perception, and the "arrow of time" is related to the perception of the instant "now." This is really not part of physics but lies in the province of human consciousness—what we call the "C-question." We have recorded memories of the events that have occurred throughout our existence, and our brain continually compares new events to these, creating a perceived interface between the future and the past, which we sense as "now." All physics questions are phrased such as the following question: "Given that 'the thing' starts at time t_1 in the position x_1, what will be its position at time t_2?" As we see, time transla-

tional symmetry means that the following question will have the same answer as the first: "given that 'the thing' starts at time $t_1 + T$ in the position x_1, what will be its position at time $t_2 + T$?"

6. We can write a formula for (x', y') in terms of (x, y) and the rotation angle, θ, using trigonometry. The result is $x' = x \cos(\theta) + y \sin(\theta)$ and $y' = -x \sin(\theta) + y \cos(\theta)$. Substituting these expressions into $L = \sqrt{x'^2 + y'^2}$, we get $L = \sqrt{x^2 + y^2}$. So our mathematical formulas give the same value of the length of the pointer *after* we perform a rotation—that is, our math contains rotational symmetry.

7. We do not recommend that you repeat this experiment at home. If one falls into the black hole, while the center of mass of the rigid system, that of the spacecraft, is in free fall while the endpoint extremities are not; they are attached rigidly to the center of mass and cannot themselves be in exact free fall. This gives rise to a stress called a tidal force, which becomes so strong near the event horizon of the black hole that it will tear the unlucky faller limb from limb. For most gravitating systems, such as the Sun or Moon or Earth, this is not a large effect, since gravity doesn't vary much over the size of the spacecraft. We do, of course, see tidal forces acting on Earth, which has a center of mass that is "free-falling" (or orbiting) about the Earth-Moon center of mass; the surface of the sea is not freely falling, however, and is free to flow by the tidal forces; these forces may be responsible for other effects, such as earthquakes.

CHAPTER 5

1. The concept of a conserved quantity is related to the concept of a *conserved current*. An example of a conserved current is the flow of electrical charge in a system, and the imaginary box might be a circuit component, such as a capacitor or a resistor. Electrical charge is never created or destroyed at any local point in an electrical circuit—that is, electrical charge is conserved—but electrical charge can flow *to* or *from* the point in an electric current, so this must be taken into account when talking about the conservation of electric charge. We end up with the following statement: "The time rate of change of the electric charge in any local imaginary box equals the current flow into or out of the imaginary box." This is another restatement of a conservation law, and electric charge conservation indeed follows from a deep and fundamental symmetry of electromagnetism, all governed by Noether's theorem.

2. As an example of vectors, weather maps often show the wind velocity at various places on the map, indicated by little vectors called "wind barbs." The weather maps usually refer to a given altitude, such as the "18,000-foot map," or to a given atmospheric pressure, such as the "500-millibar map" (these are roughly the same, varying slightly when high or low pressure systems move in). The wind velocity at a given location is usually indicated by a line segment attached to an open circle pointing in the direction the wind is blowing from, with tick marks that indicate the speed of the wind in units of 10 knots (long "barb ticks") or 5 knots (short "barb ticks"; 1 knot = 1 nautical mile per hour = 1.15 miles per hour). If the circle contains a pennant, we add 50 knots to the vector. This is just a variation on the symbolic representation of a vector—the concept is the same as a line segment with an arrow representing direction and length representing the speed.

3. It is impossible to talk meaningfully about physics without introducing the concept of a vector. Vectors are usually denoted by a symbol with an arrow on top, such as $\vec{v}$, or $\vec{u}$, or $\vec{P}$, or $\vec{p}_1$, and so forth. If we construct a coordinate system, then we say that the vector has components that are its projection along each of the three axes of the coordinate system; for example, $\vec{p}_1 = (p_x, p_y, p_z)$, where each component is the projection of the vector along the corresponding coordinate axis. Vectors can be added or subtracted, and they can be multiplied by ordinary numbers to increase or decrease their magnitudes (lengths). The total momentum of a system is the sum of all of the individual momenta of all the components of the system. We often write this as an equation, that is, $\vec{P} = \vec{p}_1 + \vec{p}_2 + \vec{p}_3 + \ldots$, where $\vec{P}$ is the total momentum of the system and $\vec{p}_1$, $\vec{p}_2$, $\vec{p}_3$, and so forth are the momenta of the individual component parts of the system. Noether's theorem states that $\vec{P}$ is conserved, whereas the individual $\vec{p}_i$ may change in a given process. Moreover, the formula $\vec{P} = m\vec{v}$ applies to an approximately pointlike massive object (and, assuming it moves at a velocity small compared to the speed of light) the momentum is just the mass of the object times its *velocity*. The directions in space in which physical systems can be translated (moved) are vectors, so, if a student remembers Noether's theorem, he won't forget that momentum is a vector when taking the SAT!

4. Readers will note that this is a slight variation of the process $p^+ + e^- \rightarrow n^0 + \nu_e$, which destroys a Titanic star and causes a supernova. The squeezing together of a proton and electron can happen only at the extreme densities inside a massive star collapse. A neutron in free space

decays into a proton, electron, and (anti-)neutrino in a half-life of about eleven minutes by the related process of "beta decay," $n^0 \rightarrow p^+ + e^- + \bar{\nu}_e$.

5. Remarkably, Earth did once take a direct hit from a pretty hefty planetoid, and the Earth-Moon system was born. Details of this theory, which fairly accurately predicts abundances of things like water, iron, silicon, etc., on the Moon, will always be somewhat limited without detailed knowledge of the initial Earth and planetoid system. See, e.g., W. Benz, A. Cameron, and H. J. Melosh, "The Origin of the Moon and the Single-Impact Hypothesis III," *Icarus* 81 (1989): 113–31; H. J. Melosh, "Giant Impacts and the Thermal State of the Early Earth," in *Origin of the Earth*, ed. H. Newsom and J. Jones (Oxford: Oxford University Press, 1990), pp. 69–83. Melosh has an informative Web site, "Origin of the Moon," http://www.lpl.arizona.edu/outreach/origin/ (accessed May 18, 2004).

6. We can understand "wobbling" by thinking of the orbital motion as though it consists of many instantaneous interactions, or "feeble collisions," of Earth and the Sun, through the force of gravity. Then the initial total momentum is $m_{\text{Earth}}\vec{v}_{\text{Earth}} + m_{\text{Sun}}\vec{v}_{\text{Sun}}$. The final total momentum is $m_{\text{Earth}}\vec{v}'_{\text{Earth}} + m_{\text{Sun}}\vec{v}'_{\text{Sun}}$. In these collisions the mass of Earth and the Sun is conserved (ignoring the Moon, Jupiter, Mars, etc.), or $m_{\text{Earth}}\vec{v}_{\text{Earth}} + m_{\text{Sun}}\vec{v}_{\text{Sun}} = m_{\text{Earth}}\vec{v}'_{\text{Earth}} + m_{\text{Sun}}\vec{v}'_{\text{Sun}}$. However, we know that Earth is much less massive than the Sun, or $m_{\text{Earth}} << m_{\text{Sun}}$. So we can now do a little algebra and we see that after the "collision," the change in the velocity of the Sun is

$$\vec{v}'_{\text{Sun}} - \vec{v}_{\text{Sun}} = \left(\frac{m_{\text{Earth}}}{m_{\text{Sun}}} \right) (\vec{v}'_{\text{Earth}} - \vec{v}_{\text{Earth}}).$$

The change in the velocity of the Sun is proportional to the tiny quantity $m_{\text{Earth}}/m_{\text{Sun}}$. For Earth and the Sun this number is extremely small, about 0.3×10^{-6}. So any change in the velocity of the Sun, or "wobbling" of the motion of the Sun due to Earth's orbit, is almost imperceptible. Jupiter is much more massive than Earth, with $m_{\text{Jupiter}}/m_{\text{Sun}} \approx 10^{-3}$, and it could obviously cause more wobbling of the Sun, but Jupiter's orbital radius is larger, and its orbital velocity in space is much less than Earth's, so the effect is reduced somewhat, to less than 10^{-3}. This, by the way, also explains why Earth doesn't recoil significantly when you jump up off the ground—Earth does indeed recoil for an instant to conserve momentum, and it undergoes a slight change in its velocity, but only by an amount equal to your mass divided by Earth's mass, times your jump velocity. This number is minuscule!

7. Astronomers can tell if a distant star is wobbling by using the Doppler effect on the emitted light (the redshift of the color of receding light sources or the blueshift as they approach). Wobble watching resulted in the discovery of the first few new exoplanets: 51 Pegasus, a planet roughly the mass of Jupiter with an orbital radius less that than the Mercury-Sun orbital radius and with an orbital period of 4.2 days. (By comparison, Mercury takes 88 days to orbit our Sun.) For more information on exoplanets visit the following Web sites: "Wobble Watching Revisited," Starryskies.com, http://www.starryskies.com/articles/dln/5-96/ newpls.html; Laurence R. Doyle, "Detecting Other Worlds: The Wobble Method," Space.com, http://www.space.com/searchforlife/seti_wobble _method_010523.html; Maya Weinstock, "Astronomers Discover Bundle of Extrasolar Planets," Space.com, http://www.space.com/scienceas-tronomy/astronomy/new_planets_000804.html (all accessed May 18, 2004).

8. The letter is in the CERN Pauli Archive, which may be visited at http://library.cern.ch/archives/pauli/paulimain.html (accessed June 1, 2004). We thank the CERN Pauli Archive Committee for permission to reproduce it. *Neutrino* occurs in brackets here because it was actually Enrico Fermi who gave the particle this name. Pauli had used the name *neutron* for his new particle, the name we now use for the heavy neutral constituent of the nucleus.

9. We have greatly simplified the discussion here. For spin, or any circular motion, it is useful to talk about the "angular velocity vector," which we usually denote by the Greek letter omega, $\vec{\omega}$. This is a vector with a magnitude that is just the number of radians (remember, 360 degrees equals 2π radians) that the object rotates through per second. The direction of $\vec{\omega}$ is defined by the right-hand rule. For a planet in a circular orbit, the magnitude of the velocity is $|\vec{v}| = |\vec{\omega}R|$, and therefore the magnitude of the momentum is $|\vec{p}| = |m\vec{\omega}R|$ (pointing tangential to the orbit), and the magnitude of the angular momentum is $|\vec{J}| = |m\vec{\omega}R^2|$, pointing out of the plane of the orbit, with the sense defined by the right-hand rule.

When any object with an "approximate radius" R of mass m spins on an axis with angular velocity $\vec{\omega}$, it will have a spin angular momentum of $\vec{J} = km\vec{\omega}R^2$, where k is a number that characterizes the shape and internal matter distribution of the object. For example, if the object is a disk, and the rotation is in the plane of the disk, then $k = 1/2$, whereas if it is a ring, $k = 1$. The value of k is determined by summing all of the circular orbital angular momentum of all the pieces (atoms) that make up the object (this involves integral calculus). We usually define the "moment of inertia" of an object to be $I = kmR^2$, which refers to the object's internal shape, size, and structure, or the "guts" of the thing. The angular momentum of a spinning object is then $\vec{J} = I\vec{\omega}$. In the three-dumbbell experiment, what is kept constant is the total angular momentum, $\vec{J} = I\vec{\omega}$. By bringing the dumbbells close to his body, the professor's moment of inertia I, which is proportional to R^2, is decreased, but $\vec{J} = I\vec{\omega}$, the angular momentum, must be conserved, so $\vec{\omega}$ must increase significantly, as $\propto 1/R^2$. This is why the demonstration is so impressive: halving R increases the angular frequency fourfold.

CHAPTER 6

1. A historical summary can be found in Will Durant and Ariel Durant, *The Story of Civilization*, vol. 7, *The Age of Reason Begins* (New York: Simon and Schuster, 1983).

2. Feynman, a codeveloper of quantum electrodynamic theory, the precise description of the electron interacting with the photon, invented "Feynman diagrams," a graphical way of organizing the complex calculations that govern the motion and interactions of matter at the quantum scale (see chap. 11). Feynman made many other significant contributions to our understanding of nature, including his own formulation of the laws of quantum mechanics, which has proven indispensable in modern developments of particle physics. Every practicing physicist knows the famous Feynman Lectures, a three-volume, college-level course in physics still completely relevant today, some forty years after they were delivered in the 1960s at Caltech. The nation briefly became acquainted with Feynman publicly when he served on the commission that investigated the space shuttle *Challenger* disaster. He graphically demonstrated, on live television, the susceptibility of the rubber O-ring material used in the solid-fuel rocket boosters to freezing and accompanying loss of flexibility, which led to the escaping of gases through the side of the rocket and ultimately to the space shuttle's disastrous fate. His dissenting report on the *Challenger* disaster hauntingly anticipated the *Columbia* disaster seventeen years later.

3. Richard P. Feynman, as told to Ralph Leighton, *What Do You Care What Other People Think?* (New York: W. W. Norton, 1988), p. 15.

4. It was as if the Greeks had an equation of motion in mind: force

equals mass times velocity, or $\vec{F} = m\vec{v}$. This would mean that to move an object with any finite velocity, one would need to apply a force. If the object were heavier, one would require a greater force. The motion of the object would always be in the direction of the applied force. We emphasize that this is *not* the correct equation of force and motion! It is Newton's equation $\vec{F} = m\vec{a}$, where $\vec{a}$ is the acceleration, or change in velocity per unit time, that is the correct equation.

5. Arthur Koestler, *The Sleepwalkers: A History of Man's Changing View of the Universe* (London: Penguin Press, 1959). Koestler was a champion of Kepler, whom he saw as the central player in the story of the unfolding of physics.

6. Throughout the Middle Ages monastic scholars fine-tuned Ptolemy's theory to obtain precise predictions. They were unwittingly inventing modern "Fourier analysis," by which any mathematical function can be approximated by adding together a series of periodic (trigonometric) functions. See Emmanuel Paschos, *The Schemata of the Stars: Byzantine Astronomy from 1300 A.D.* (Singapore: World Scientific Press, 1998).

7. There will be some "wiseguys" who object to this strong statement. For example, for the case of one object orbiting another of the same mass, it is confusing to say that one orbits the other—they both orbit each other. On the other hand, we can actually use a moving coordinate system attached to either object and describe the motion in terms of that coordinate system, which then treats the other object as if it is orbiting. Thus, technically we *can* say the Sun orbits Earth, in an Earth-bound coordinate system (Einstein's general relativity certainly permits the use of any coordinate system we choose), but we certainly cannot then say that the other planets, Venus, Mars, etc., also orbit Earth, since they won't do so in that Earth-bound coordinate system. To be precise, in an inertial reference frame, all objects orbit the center of mass of the solar system, which can be viewed as fixed in space. The center of mass is approximately located at the center of the Sun, because the Sun is so massive.

8. The scripture was later confronted explicitly by Galileo; see, e.g., Dava Sobel, *Galileo's Daughter* (London: Fourth Estate, 1999).

9. For an excellent article about the historical research of Owen Gingerich, see Christopher Reed, "The Copernicus Quest," *Harvard Magazine*, November 2003; see also Owen Gingerich, *The Book Nobody Read: Chasing the Revolutions of Nicolaus Copernicus* (New York: Walker, 2004).

10. Kepler's laws, stated more succinctly, are as follows: (1) the orbits of the planets are ellipses, with the Sun at one focus of the ellipse; (2) a line from the planet to the Sun sweeps out equal areas in equal intervals of time (this is equivalent to the statement of the law of conservation of angular momentum); and (3) $T^2 = kR^3$, where T is the orbital period (in years), R is the semimajor axis of the ellipse (in astronomical units), and the constant k is the same for all planets in the solar system.

There are many Web sites, some featuring animations, that illustrate Kepler's laws. Type "Kepler's laws" into the Google search engine (http://www.google.com), or see, for example, http://www.phy.ntnu.edu.tw/java/Kepler or Bill Drennon, "Kepler's Laws with Animation," http://www.cvc.org/science/kepler.htm (both accessed June 2, 2004).

11. See Koestler, *The Sleepwalkers*, pp. 446–48.

12. An account of Newton and his contemporaries can be found in Durant and Durant, *The Story of Civilization*, vol. 7, *The Age of Reason Begins*, and vol. 8, *The Age of Louis XIV* (New York: Simon and Schuster, 1983).

13. The key here is to determine the distance, x, that is covered by an object accelerating at a rate a in time t. The formula is easy to derive after the first week in a calculus class, but we'll give you the answer: $x = \frac{1}{2} at^2$. Now, the distance from New York City to Chicago we'll take, for the sake of discussion, to be 1200 km (about 800 miles). Therefore the distance to the halfway point is 600 km, and the time this part of the trip takes is determined by our formula: 600 km = 600,000 m = $\frac{1}{2} \times 5$ ms$^2 \times t^2$, and we find that $t = 490$ seconds. Since it takes the same amount of time to decelerate and arrive in New York City (by symmetry, we can just imagine running the clock backward in this phase of the motion), therefore, the entire trip takes $T = 2t = 980$ seconds, or only about 16.3 minutes!

14. To get a feeling for the feebleness of the gravitational force, pick up a full gallon container of milk. The force you are exerting to do this is a little more than eight pounds. This is the approximate strength of the gravitational force of attraction between two completely filled oil tankers ten miles apart.

15. In fact, 2½ centuries after Newton, Ernest Rutherford was launching electrically charged alpha particles (as discovered earlier by the Curies) at the atom and found that exactly the same scattering trajectories occur there, which established that atoms have a Sun-like center called the nucleus. The alpha particles were scattered by the electromagnetic force from the nucleus, which is an inverse square law force, like gravity.

16. The discovery of the planetoid Sedna was announced in March 2004. It is the tenth planet in our solar system, with an extremely elliptical orbit, but it is believed that many other similar, distant objects exist. See, e.g., Michael E. Brown, "Sedna (2003 VB12)," http://www.gps .caltech.edu/~mbrown/sedna/; "Sedna (planetoid)," Wikipedia, http://en .wikipedia.org/wiki/Sedna_(astronomical_object) (both accessed June 3, 2004).

CHAPTER 7

1. Professor Michael Fowler, of the University of Virginia, provides an excellent Web site on the history and physics of relativity, including the measurement of the speed of light; see Fowler, "Galileo and Einstein," http://galileoandeinstein.physics.virginia.edu/ (accessed June 3, 2004).

2. Dava Sobel, *Longitude* (New York: Walker, 1995), chronicles the fascinating account of the general history of the problem and solution of longitude. The astronomers actually obstructed awarding a prize for the heroic, single-minded efforts of John Harrison, who constructed the first seaworthy clocks.

3. See ibid., pp. 11–13, for details on Admiral Sir Clowdisley Shovell's tragic foundering on the Scilly Isles, off the southwest tip of England.

4. The Galilean moons can be seen with any inexpensive backyard telescope on a clear night as Jupiter hovers conspicuously in the sky. Jupiter and its moons are a system structured much like the solar system but in miniature. The orbits of the moons are approximately circular, and the orbital periods and motions are determined by Kepler's laws, which are governed in turn by Newton's universal law of gravitation and the principle of inertia. The moons have now been photographed in detail by the NASA/Jet Propulsion Laboratory (JPL) fly-by satellite *Galileo*; see "Galileo: Journey to Jupiter," http://www2.jpl.nasa.gov/galileo (accessed June 3, 2004).

5. Assume circular orbits. Let T_E be Earth's orbital period (one year) and T_M be Mars's orbital period (1.88 years). And let R_E be Earth's orbital radius (the AU, or astronomical unit, which we seek) and R_M the Martian orbital radius. At opposition (i.e., the closest approach of Mars to Earth) we have $R_M = R_E + d$, where d is the distance measured by Cassini using

parallax with the naval vessel in the South Pacific. Using Kepler's third law we have $(T_M/T_E)^2 = (R_M/R_E)^3$. Therefore, substituting and solving, we get $R_E = d/[(T_M/T_E)^{2/3} - 1] = 1.91d$. Unfortunately, it isn't quite as simple as this because the Martian orbit is very elliptical, and d can vary between 35 million miles (56 million km) and about 63 million miles (100 million km). Kepler's law refers to the length of the semimajor axis of the elliptical orbit, which Cassini had to work out from his data. The correct answer is obtained by using the average of the least distance at closest opposition (such as we experienced in the year 2003) and the maximum, a value of 49 million miles. This gives $R_E = 1.91 \times 49 = 92$ million miles (148 million km).

6. To simplify any discussion involving the two different coordinate systems of two different observers, we'll agree to use the same units of measurement. We also agree, at least at the outset, to have the two different coordinate systems with axes that are *parallel*. This means that we agree on the direction of the x-axis, the y-axis, and z-axis, and both observers use these conventions. When we talk about time we may also want our clocks to be synchronized. Also, when we talk about motion, sometimes we'll agree on having our coordinate systems be exactly identical at a special moment in time, let's say $t = 0$, when a "nonmoving" observer (stationary observer) and a "moving" observer are located at the same place in space, the origin. To arrange all of this just involves time and space translations, which are symmetries of physics. We don't have to do this "calibration," but it will often be useful. The coordinate system of the moving observer will move with the moving observer, while the nonmoving observer's coordinate system will stay at rest.

7. More generally, given two coordinate systems, the Galilean boosts are $x' = x - vt$, $y' = y$, $z' = z$, and $t' = t$ for motion along the positive x-axis. Since this is a continuous symmetry transformation, there is a corresponding conservation law that is not hard to determine; see E. L. Hill, "Hamilton's Theorem and the Conservation Theorems of Mathematical Physics," *Review of Modern Physics* 23 (1953): 253. There are three directions in space in which we can boost an object; hence, the conserved quantity must be a vector. Consider a system containing a large number of masses, m_a. We define the *mass-position* of the system, $\vec{Q} = \sum m_a \vec{r}_a$, which is the sum of the position vector of each particle multiplied by its mass (this quantity is related to what we call the center of mass of the system, which is $\vec{X} = (\sum m_a \vec{r}_a)/(\sum m_a) = \vec{Q}/M$, where $M = \sum m_a$ is the total mass of the system). Also, the total momentum of the system is

$\vec{P} = \sum m_a \vec{v}_a = \dot{\vec{Q}}$, which is conserved, $\dot{\vec{P}} = 0$, by translational invariance. Then, the Noetherian conserved quantity corresponding to boosts is the vector $\vec{K} = \vec{Q} - \vec{P}t$, which is easily seen to be conserved, $\dot{\vec{K}} = 0$. This implies that the center of mass of a system moves through space with a constant velocity. A more thorough derivation leads to the conclusion that the *center of mass* of a system will move with constant velocity for any internal physical process.

8. A. Einstein, "On the Electrodynamics of Moving Bodies," *Annalen der Physik* 17 (1905): 891–921 [in German]; reprinted in the collection *The Principle of Relativity* (New York: Dover, 1952), pp. 35–65. The mysterious role played by Einstein's first wife, Mileva Marić, in the development of special relativity has recently come under scrutiny. Her tragic fate somewhat tarnishes the docile and grandfatherly image of the great man. See "Einstein's Wife: The Life of Mileva Marić Einstein," PBS Web site, http://www.pbs.org/opb/einsteinswife/ (accessed June 4, 2004), and resources listed there.

9. This is a simplified version of the Einstein boost, not the most general form. The moving observer could be moving in any direction relative to the two events. Also, we have been talking about length and time intervals, to avoid the longer discussion of coordinate systems, but coordinate systems provide a more general language to phrase the results. A stationary observer would describe events as points in space-time labeled by four coordinates, (x,y,z,t). A moving observer carries a "comoving" coordinate system (x',y',z',t'). The Einstein boosts relate these for relative velocity in the $+x$ direction with speed v:

$$x' = \gamma (x - vt), \quad y' = y, \quad z' = z, \quad t' = \gamma (t - vx/c^2),$$

$$\text{where } \gamma = \frac{1}{\sqrt{1 - v^2/c^2}} \ .$$

These replace the Galilean boosts in note 7 above.

10. We can easily check with a little algebra that the invariant interval is the same for two observers:

$$\tau^2 = T'^2 - L'^2/c^2 = \gamma^2 (T - vL/c^2)^2 - [\gamma^2 (L - vT)^2]/c^2 = T^2 - L^2/c^2,$$

$$\text{where } \gamma = \frac{1}{\sqrt{1 - v^2/c^2}} \ .$$

Hence, the interval, or proper time, is invariant under boosts. Stated in coordinate language, given any two events 1 and 2, the interval between the events is

$$\tau^2 = (t_1 - t_2)^2 - [(x_1 - x_2)^2 + (y_1 - y_2)^2 + (z_1 - z_2)^2]/c^2,$$

where c is the speed of light. A moving observer would write

$$\tau^2 = (t'_1 - t'_2)^2 - [(x'_1 - x'_2)^2 + (y'_1 - y'_2)^2 + (z'_1 - z'_2)^2]/c^2.$$

The Lorentz transformation preserves the invariant interval, since

$$\tau^2 = t'^2 - [x'^2 + y'^2 + z'^2]/c^2 = \gamma^2(t - Vx/c^2)^2 - [\gamma^2(x - Vt)^2 + y^2 + z^2]/c^2$$
$$= t^2 - [x^2 + y^2 + z^2]/c^2.$$

In the language of groups (see the appendix), the symmetry is similar to $SO(4)$, the symmetry of the four-dimensional sphere. An ordinary rotation in the x-y plane mixes the x and y coordinates with factors such as $\cos(\theta)$ and $\sin(\theta)$. Lorentz transformations along the x-axis are like rotations that mix x and ct, with factors γ and $-\gamma V/c$. Note that while $\cos^2(\theta) + \sin^2(\theta) = 1$, we now have $\gamma^2 - (-\gamma V/c)^2 = 1$. The difference between Lorentz transformations and ordinary rotations in four dimensions is the presence of the minus signs in the proper time that distinguish time from space. We define the symmetry group to be $SO(1,3)$. That is, while $SO(4)$ is the collection of transformations on four coordinates (x,y,z,w) that leave invariant the radius of the unit sphere $1 = x^2 + y^2 + z^2 + w^2$, $SO(1,3)$ is the collection of transformations on the four coordinates that leave invariant the quantity $1 = -x^2 - y^2 - z^2 + w^2$. This is a continuous symmetry group called the *Lorentz group*.

11. To understand length contraction, let us ask what the moving observers would measure for the distance between events in their reference frame. They would indeed measure the length interval L' and time interval T' between two events, where $L' = \gamma \cdot (L - vT)$ and $T' = \gamma \cdot (T - vL/c^2)$. However, when we measure the *length of an object* we must measure the distance between two events located at the endpoints that are *simultaneous* to us, so the moving observers insist that $T' = 0$. Therefore $T = vL/c^2$ and $L' = \gamma \cdot (L - vT) = \gamma \cdot (L - [v^2/c^2]L)$, which is $L' = \sqrt{1 - v^2/c^2}\, L$. Time dilation is easy, since the metronome is flashing with time interval T and $L = 0$. So the moving observers measure $T' = \gamma \cdot (T - vL/c^2) = \gamma \cdot T$, and the gamma factor is the time dilation.

We can now understand why I cannot catch Ollie the cat when he is traveling at the speed of light. Ollie begins to run at space-time event 1 with (x,y,z,t) coordinates $(0,0,0,0)$ in the $+x$ direction with a speed u. At a later time T, Ollie passes another point in space, defining space-time event 2, with coordinates $(uT,0,0,T)$. Ollie's speed, in the stationary frame, is just the difference in spatial coordinates, $uT - 0 = uT$, divided by the difference in time coordinates, $T - 0 = T$; hence, $uT/T = u$.

Now, we suppose that I run in the $+x$ direction at velocity $+v$ relative to the stationary frame. What do I measure for Ollie's velocity? From the Lorentz transformation in note 9 above to my moving coordinates (x',y',z',t'), event 1 is $(0,0,0,0)$, and event 2 is $(\gamma(uT - vT),0,0,\gamma(T - u vT/c^2))$. I therefore find the x distance between these events, $\gamma(u - v)T$, and the time interval is $\gamma(1 - uv/c^2)T$, so I obtain the speed, in the usual way, as the ratio of distance over time interval: $u' = \gamma(u - v)T/\gamma(1 - uv/c^2)T$; hence, we obtain $u' = (u - v)/(1 - uv/c^2)$. Thus u' is the speed that I observe Ollie to have, running away from me as I chase him with speed v.

This formula is called the *velocity addition formula*, for the special case of parallel motion. Notice that when we take the speed of light to be infinite, it reduces back to $u' = (u - v)$, which was exactly what Galilean physics would predict. If Ollie is traveling at the speed of light, then $u = c$, and I will measure Ollie's speed to be $u' = (c - v)/(1 - cv/c^2) = c$! No matter what my velocity v is, I will always measure Ollie, or light waves, receding away from me at the same velocity, c. Of course, this merely reaffirms the starting point of special relativity, which built the constancy of the speed of light into the theory ab initio.

12. This result implies that one observer will see that a particle has an energy and momentum given by $(E, \vec{p})$, and another observer moving relative to him, with a velocity $\vec{v}$ in the $+x$ direction, will observe the same particle to have a different energy and momentum, $(E', \vec{p}')$. However, these will be also related by the Lorentz transformation:

$$p_x' = \gamma(p_x - vE/c^2), \quad p_y' = p_y, \quad p_z' = p_z, \quad E' = \gamma(E - vp_x).$$

The moving observer will find that, despite the fact that the energy and momentum are now changed, the inertial mass is still the same:

$$E'^2 - |\vec{p}'|^2 c^2 = m^2 c^4.$$

13. This comes from the Taylor series approximation to a square root,

$\sqrt{a^2 + x^2} \approx a + x^2/2a$ for $x \ll a$. We can turn this around to get the final formulae for the energy and momentum of a moving particle. Assume that the particle is at rest in the rest frame. Then the energy and momentum are ($E = mc^2$, $\vec{p} = 0$). Now, boost these to a frame in which the particle is moving with a velocity of $\vec{v} = (v,0,0)$ (note that this is a boost of $-\vec{v}$). Then we find the energy and momentum of the moving particle:

$$\text{Einstein: } E = \frac{mc^2}{\sqrt{1 - v^2/c^2}}, \vec{p} = \frac{m\vec{v}}{\sqrt{1 - v^2/c^2}};$$

$$\text{Newton: } E = \frac{1}{2}mv^2, \vec{p} = m\vec{v}.$$

We have included the Newtonian expressions for comparison. This is a whopping big difference. Again, we see the rest energy implied for $\vec{v} = 0$, where the Einstein formulae go into

$$E \approx mc^2 + \frac{1}{2}mv^2, \text{ and } \vec{p} = m\vec{v}.$$

We see that in special relativity, we can never get the speed of a massive particle (one with nonzero inertial mass m) to equal the speed of light. As $|\vec{v}| \to c$, the momentum and the energy become infinite. It therefore would require an infinite energy to accelerate a proton to the speed of light. At the Fermilab Tevatron we accelerate protons to one trillion electron volts of energy. The rest mass energy of a proton is about one billion electron volts. Hence, the Tevatron boosts a proton to have a Lorentz factor, $\gamma = 1/\sqrt{1 - v^2/c^2}$, of about 1,000. This means that $v/c \approx 0.9999995$, or that the Tevatron accelerates protons to 99.99995 percent of the speed of light. The protons never quite get to the speed of light, even at the highest-energy accelerator in the world! How, then, can anything travel at the speed of light? We see that if we take $|\vec{v}| = c$ and allow our particle to be massless, then the energy is actually indeterminate; that is, we get $E = 0/0$. However, this allows for the possibility that a massless particle, something with no inertial mass, can have finite energy and momentum. If we look at the relationship between energy and momentum, we see that a massless particle must satisfy $E = |\vec{p}|c$. Indeed, this describes the particles of light, the photons. Photons have absolutely no inertial mass, yet they transmit energy and momentum through space. Photons travel forever at the speed of light. They cannot be at rest or have a finite velocity less than c, for then their energy would be zero.

14. Since a particle at rest has an energy that is equivalent to its inertial mass, we can therefore measure a particle's mass in terms of this energy. It is convenient to use an energy unit other than joules to do this. In particular, we use a very tiny quantity of energy, the *electron volt* (abbreviated eV), which is the energy that a one-volt battery expends pushing a single electron through a circuit. The conversion shows how tiny this is: 1 joule = $6.24150974 \times 10^{18}$ eV. However, it is useful—in fact, the proton mass, which is $1.67262158 \times 10^{-27}$ kilograms, can be stated in terms of electron volts as $m_{proton}c^2 = 1.5 \times 10^{-10}$ joules = 938 MeV (1 MeV means one million eV, or 10^6 eV). We often make rough estimates, taking the proton and neutron masses to be approximately 1 GeV (meaning one billion eV, or 10^9 eV).

If we burn carbon, combining a carbon atom, C, and an oxygen molecule, O_2, we produce CO_2, getting out about $E = 10$ eV in energy (in photons). The mass of the resulting CO_2 molecule is therefore actually less than the mass of the initial C and O_2 by the tiny amount E/c^2. This is a fractional decrease in mass of the carbon + oxygen molecule (with about 12 + 16 + 16 protons and neutrons, or 46 GeV of mass) of about 10 eV/46 GeV $\approx 0.2 \times 10^{-9}$. This represents the conversion efficiency, so, to meet the US energy needs, we need to burn 1,000 kg/(0.2×10^{-9}) $\approx 5 \times 10^{12}$ kilograms of oil per year. In nuclear fission, a uranium-235 nucleus will typically convert to lighter nuclei (such as yielding about 200 MeV per fission. This is a conversion efficiency of 200 MeV/ (235×1 GeV) $\approx 10^{-3}$, much more efficient than burning carbon. In nuclear fusion, we can combine a hydrogen nucleus (proton) with a deuterium nucleus (proton + neutron) to produce a helium isotope (2 protons and one neutron) with a release of 14 MeV of energy, therefore a conversion efficiency of about 4 $\times 10^{-3}$.

15. Some good introductory books on general relativity include Robert M. Wald, *Space, Time, and Gravity: The Theory of the Big Bang and Black Holes* (Chicago: University of Chicago Press, 1992); Clifford Will, *Was Einstein Right?* (New York: Basic Books, 1993); for the more advanced student, the best one is Steven Weinberg, *Gravitation and Cosmology* (New York: John Wiley and Sons, 1972).

16. Technically, the condition that all of the energy of a particle of mass, m, would have to be expended in order to escape from the pull of the large object, such as a star, of mass M with a radius of R, is that the particle's rest energy, mc^2, must equal or exceed the gravitational potential energy that traps it, which by Newton's theory is $mc^2 = G_N Mm/R$. This

predicts that the *Schwarzchild radius* is $R = [2]G_N M/c^2$ without the factor of [2]. We have inserted this factor into this formula, which is what is found when the calculation is done correctly in general relativity.

CHAPTER 8

1. T. D. Lee, and C. N. Yang, "Question of Parity Conservation in Weak Interactions," *Physical Review* 104 (1956); J. Bernstein, "Profiles: A Question of Parity," *New Yorker Magazine* 38 (1962); M. Gardner, *The New Ambidextrous Universe: Symmetry and Asymmetry, from Mirror Reflections to Superstrings* (New York: W. H. Freeman, 1991).

2. For amusing anecdotes surrounding the discovery of parity violation in the decay of the pion, and well as in the decay of the muon, see Leon M. Lederman, *The God Particle* (New York: Dell, 1993).

3. In other words, if we view the system in a particular mirror, we can see that the magnetic field will appear to reverse direction, whereas the direction of motion of the outgoing electrons will remain the same; with a different position of the mirror, the motion of the electrons will reverse, while the magnetic field stays the same. The relative alignment of motion and field always reverses.

4. As an aside, notice yet another aspect of the kinds of questions we ask in physics. Nowhere in any formulation of physics that we know of does the issue of a special point in time called "now" ever occur. Yet we humans sense something we call "now." Is "now" an illusion? Since this is a puzzling question, we give it an air of profundity and call this the "N question." Special relativity tells us that there is no absolute "now" in the universe, because different observers in different inertial reference frames will disagree on which events at different places in space are simultaneous. Hence, even within our brains, there can be no perfect synchronization on extremely short timescales on the order of the size of our brain divided by the speed of light. However, the brain is fairly slow, since interneural communication takes about a millisecond to happen, so perhaps there is some time-averaging going on that yields the experience of "now." Is "now" therefore real and part of the laws of physics? The fact that this question is so murky may be telling us the answer: there is no privileged role of "now" in the laws of physics. The perception of the sensation of "now" has to do with the murky business of "consciousness" (which we previously called the "C question"). Since there is as yet no

complete theory (that we are aware of), or even a good predictive model, of consciousness, we cannot address this further, except to say that consciousness is a very complex phenomenon. We suspect the answer is that "N" and "C" are related to each other.

5. The K-meson particle, K^0, and the antiparticle, $\bar{K}^0$, actually oscillate back and forth between one another, $K^0 \leftrightarrow \bar{K}^0$. If CP were an exact symmetry, then the oscillation phase from K^0 into $\bar{K}^0$ should be *exactly the same* as the reverse oscillation phase, from $\bar{K}^0$ into K^0. Experimentally, however, it is found that the oscillation phase from $K^0 \to \bar{K}^0$ is slightly different, at one part in a thousand, than the oscillation phase from $\bar{K}^0 \to K^0$; see J. H. Christenson et al., "Evidence for the 2 Pi Decay of the K^0 Meson," *Physical Review Letters* 13, nos. 138–40 (1964). This is not CP invariant. In refined experiments with neutral K-mesons, direct confirmation of the violation of T symmetry has also been confirmed. The combined symmetry transformation, CPT, is a symmetry of the decays. The violation of CP is now turning up in other particles, called B-mesons, containing the heavy beauty quark.

CHAPTER 9

1. This is called a "Z_N" discrete symmetry if there are N little girls.

2. See Paul Doherty, "2,000 Years of Magnetism in 40 Minutes," Technorama Forum Lecture, http://www.exo.net/~pauld/technorama/technoramaforum.html (accessed June 8, 2004).

3. The "northward"-pointing pole of a compass needle magnet, called the "north pole" of the magnet, aligns in the direction of the "north magnetic pole" of Earth, which is actually, therefore, the "south magnetic pole" of Earth, considered as a magnet!

4. Robert L. Park, America's Strange Attraction: Magnet Therapy for Pain," *Washington Post*, September 8, 1999; see also "Magnet Therapy: What's the Attraction?" *Science Daily*, September 9, 1999, www.sciencedaily.com/releases/1999/09/990909071842.htm (accessed June 8, 2004).

5. See Park, "America's Strange Attraction"; Robert L. Park, *Voodoo Science* (Oxford: Oxford University Press, 2000); see also the American Institute of Physics collection of weekly installments by Robert Park, entitled "What's New?" www.aps.org/WN/index.cfm (accessed June 8, 2004).

CHAPTER 10

1. If x is the position along the direction of motion of the wave and t is the time, then we can describe a particular traveling water wave by a sinusoidal function of the form $\psi(\vec{x}, t) = A \cos(kx - \omega t)$. When plotted at any time t, this is a *wave train*, and as the time t increases, the wave train moves to the right. The quantity k is called the *wave number*, and the quantity ω is called the *angular frequency* of the wave. These things are often related to the usual "cycle-per-second" *frequency*, $f = \omega/2\pi$, and the *wavelength*, $\lambda = 2\pi/k$. The wavelength (λ) is the distance between two neighboring troughs or crests of the wave. The frequency (f) is the number of times per second that the wave undulates up and down through complete cycles at any fixed position x. In other words, if you think of the wave as a long freight train, then λ is the length of a boxcar and f is the number of boxcars per second passing in front of you as you patiently wait for the train to pass. A is called the *amplitude* of the wave and determines the height of the crests; the distance from a trough to a crest is $2A$. The velocity of the traveling wave is $c = \lambda f = \omega/k$. This is usually written as a vector, and kx is usually written as $\vec{k} \cdot \vec{x}$ in three dimensions of space, to represent a wave traveling in the direction $\vec{k}$.

2. Physicists, such as Max Planck, preferred to talk about an ideal "black body," which is a cavity surrounded by a hot wall. The cavity is brought to a certain temperature, and one looks only at the light that is contained or emitted from inside the cavity. This removes uncertainties from the chemical composition of, to use our example again, dying campfire coals.

3. We tend to use the quantity $\hbar = h/2\pi$ more frequently in physics, and we often interchangeably refer to both h and $\hbar = h/2\pi$ as "Planck's constant."

4. Rutherford was targeting alpha particles (later discovered to be the nuclei of helium atoms) radiated from certain radioactive materials at thin foils of gold. The picture he had in mind was that of shooting bullets through a large blob of shaving cream. He was astonished to observe that, every now and then, an alpha particle was reflected backward, like a bullet reflected backward from the blob of shaving cream—strongly indicating something lurking inside. Rutherford found that the pattern of alpha particle scattering was exactly as expected if there were small, hard components at the centers of the atoms with positive electric charges. In this way Rutherford discovered the atomic nucleus.

5. The uncertainty principle implies that if we try to localize any particle in space within a very small region of distance, Δx, then the uncertainty in the x-component of the momentum of the particle, Δp_x, will grow larger, becoming at least as big as $\Delta p_x \leq h/2\pi\Delta x$. Similarly, if we want to localize some event in a system within a tiny time interval, Δt, then we will necessarily disturb the system and cause a range of uncertainty in its energy of ΔE, where $\Delta E \Delta t \geq h/2\pi$, so the smaller we make Δt, the larger becomes ΔE, as $\Delta E \geq \hbar/\Delta t$. The atomic orbitals of electrons have a typical size in most atoms of roughly $\Delta x \approx 10^{-10}$ meters in any given direction in space. Therefore electrons must, by the Heisenberg uncertainty principle, have a range of momentum within their orbitals that is as large as $\Delta p_x \geq \hbar/\Delta x$, hence $\Delta p_x \approx 10^{-24}$ kilogram-meters per second. Electrons move in their orbitals with velocities that are much less than c (i.e., they are *nonrelativistic*), and the electron mass is known to be $m_e \approx 9.1 \times 10^{-31}$ kilograms. Therefore we can estimate the typical electron kinetic energies to be of the order of $E \approx (\Delta p_x)^2/2m_e \approx 6 \times 10^{-19}$ joules, or about 3.8 electron volts (1 electron volt = 1.6×10^{-19} joules; we have done a lot of rounding off to do this "back of the envelope" estimate). The force that holds the electrons in their orbitals must therefore provide a negative potential energy that exceeds, in magnitude, this result. This is provided by the electromagnetic force, and the typical scale of the *binding energies* of electrons in an atom (the energy we must supply to liberate them) is of this order, ranging over about 0.1 to 10 electron volts. In fact, this is the typical energy scale of all chemical processes and contains the typical energies of visible-light photons.

6. To demonstrate this, you might wish to try the following little experiment at home or in a classroom. Put a blindfold on a subject who is seated at a table. On the table place several small objects, e.g., a pencil, a screwdriver, a quarter, a strawberry, etc. Now give the subject a balloon and instruct her to touch the objects holding only the balloon and try to discern what they are. What are their shapes? How many objects are there? By holding only the balloon, and using it as a probe of small objects, it is very difficult or impossible to answer these questions. Now give the subject a chopstick or a long piece of straw from a broom. Touching the objects with this small and refined probe, the subject can, with a little imagination and logic, reconstruct a mental picture of an object and make a hypothesis as to what it is.

7. For additional information on Schrödinger, see J. J. O'Connor and E. F. Robertson, "Erwin Rudolf Josef Alexander Schrödinger," www-

gap.dcs.st-and.ac.uk/~history/Mathematicians/Schrodinger.html (accessed June 10, 2004).

8. Here a short digression on numbers is necessary. The real numbers were discovered by the Greeks. It may seem peculiar that numbers must be "discovered," but in fact they do. We start with simple counting numbers, or the integers, 0, 1, 2, 3, etc., which were discovered by counting sheep and money, and soon we discover the negative integers, $-1, -2, -3$, etc. This discovery happened when somebody "invented" subtraction and tried to subtract 4 from 3. The Pythagoreans also invented division and discovered the *rational numbers*, that is, those numbers that can be written as the ratio of two integers, such as 3/4, or 9/28, etc. The Pythagoreans also discovered the *prime numbers*, that is, any integer that cannot be evenly divided by an integer other than itself, such as 2, 3, 5, 7, 11, 13, 17, etc. Hence, $15 = 3 \times 5$ is not prime but contains the *prime factors* 3 and 5. In a sense, the primes are the "atoms" out of which all integers are built up by multiplication. The primes are of profound importance in mathematics and remain the focus of many ongoing studies of their properties even today. Pythagoras himself did not accept the notion that there may be other numbers that cannot be written as the ratio of integers. However, numbers such as $\sqrt{2}$ and π are *irrational numbers* and cannot be expressed as the ratio of two integers. It is very hard to prove that π is irrational but rather easy to prove that $\sqrt{2}$ is irrational (Euclid himself gave the proof). These "proofs" can be found on the Internet. Taken together, the positive and negative integers, rationals, and irrationals make up the real numbers. The continuous-number line thus has a remarkable structure.

Mathematicians later discovered the *imaginary numbers*. For example, we may want the solution to the problem $x^2 = -9$. There is no real number that solves this equation. We therefore invent a new number, called i, which is defined as $i = \sqrt{-1}$. There are thus two solutions to our equation, $x = 3i$ and $x = -3i$. We can then build numbers of the form $z = a + bi$, where a and b are both real. These are called complex numbers. We define the *complex conjugate* of z to be $z^* = a - bi$ and the *magnitude* of z to be $|z| = |\sqrt{zz^*}| = |\sqrt{a^2 + b^2}|$. The imaginary numbers represent a second dimension, or a perpendicular axis, to the conventional real number line. This leads to the complex plane, in which the x-axis is the ordinary, real number line and the y-axis is the set of all real numbers multiplied by i. Complex numbers are vectors in the complex plane. A theorem of fundamental importance connects exponentials of imaginary numbers to complex numbers through trigonometric functions: $e^{i\theta}$ = $\cos(\theta) + i\sin(\theta)$. The proof of this is often relegated to a course in calculus, using Taylor series, but it can in fact be proven using just the general properties of exponentials and the "addition theorems" for trig functions (try it!). Using this result, any complex number can be written as $z = \rho e^{i\theta}$, where ρ and θ are real. Then $|z| = |\sqrt{zz^*}| = |\rho|$. This is a polar-coordinate representation of the complex plane.

9. At this point many students say, "Surely you jest! Don't you mean that you are merely using complex numbers as a kind of mathematical tool, or convenience, like people do in electrical engineering, and there really is no physical significance to the use of the complex numbers in physics equations?" To which we answer, no! We do not jest! In quantum mechanics there *really are* complex numbers, and the wave function *really is* a complex-valued function of space-time. Of course, we could reduce everything to pairs of real numbers and painfully do all the math without ever talking about the combination of the square root of −1, i, but there is no advantage to doing so. That would be like talking in a circumspect way about a horrible social disease at a cocktail party without ever using the actual name of the disease, but everyone would still understand what we were really talking about, and someone might sooner or later blurt it out. The fact is, in the mathematics of quantum mechanics the square root of −1, i, plays a *fundamental* role. Nature evidently reads books on complex numbers! We don't know why, but we know it is true. So what does a quantum particle wave function look like? Using Schrödinger's wave equation, we would find that a freely traveling particle is a wave with a wave function that looks something like this: $\psi(\vec{x},t)$ = $A(\cos(\vec{k} \cdot \vec{x} - \omega t) + i\sin(\vec{k} \cdot \vec{x} - \omega t))$, where $|\vec{k}| = 2\pi/\lambda$ and $\omega = 2\pi f$.

10. For additional information on Max Born, see J. J. O'Connor and E. F. Robertson, "Max Born," http://www-gap.dcs.st-and.ac.uk/~history /Mathematicians/Born.html (accessed June 10, 2004). See also the Olivia Newton-John international fan club biographical page, "Only Olivia," http://www.onlyolivia.com/aboutonj/index.html (accessed July 14, 2004).

11. There is actually a technically tricky point here. We really mean here the *magnitude* of the momentum, because the trapped wave is not in a state with a definite momentum, as is a traveling wave (in the language of quantum mechanics, the traveling plane wave is a momentum eigenstate, whereas the trapped wave is not). The standing wave of the guitar string has two values of the momentum at any instant, one positive and one negative, but otherwise with a definite common magnitude. The wave

function for the lowest mode is just the shape of the vibrating guitar string in space, oscillating in time. The shape of the lowest mode is the mathematical function sin $(\pi x/L)$. The exact form of the wave function involves, as it must, complex numbers, and it can be written as $\varphi(x,t) = A\sin(\pi x/L)e^{i\omega t}$, where $\omega = 2\pi E/h$. The probability of finding the electron somewhere between $x = 0$ and $x = L$ is therefore $|\varphi(x,t)|^2 = A^2\sin^2(\pi x/L)$. In fact, since the probability is unity to have the electron somewhere in the interval $0 \leq x \leq L$, then we find that $A = 1/\sqrt{2L}$.

12. A spectrometer can be constructed in half an hour with a shoe box, a diffraction grating to view through (this is a plastic grating that can be obtained for a buck at a science store or a good hobby shop, and most science teachers have them by the hundreds in that mysterious closet in the back of the chemistry or physics classroom), and a little aluminum foil. In the foil, using a razor blade or hobby knife, we cut a narrow slit to admit the light into one end of the box. At the other end of the shoe box we cut a viewing hole and attach the diffraction grating. We now close the box so that the interior viewing stage is dark. Pointing the slit at a nearby sodium vapor street light and viewing the slit by looking into the box through the grating, we will see to the side of the slit multiple copy images of the slit in rainbow colors. This will be the splayed-out spectrum of light, displaying the discrete spectral lines of emitted photons from electrons in the sodium vapor. Now let's try a more impressive target. Pointing the slit *carefully* at the Sun, one sees by looking into the box through the grating, again, the side images of the slit forming a continuous rainbow spectrum of light (but never look directly at the Sun). But looking closer we see dark lines in the rainbow. These are the photon absorption lines of hydrogen gas in the Sun's coronasphere. This phenomenon was discovered in the mid-1800s and was completely dumbfounding to physicists until the invention of the quantum theory.

13. Well, that's what science says. An alternative explanation of pulsars is that they are the communication beacons of some large, interstellar cell phone network of an advanced extraterrestrial civilization, sending messages to Communists who want to fluoridate our water. In fact, the first pulsar discovery really threw for a loop the astronomers who discovered it.

14. For more on quantum mechanics, we think the best place to start is R. P. Feynman, *The Feynman Lectures on Physics*, vol. 3 (Reading, MA: Addison-Wesley, 1963).

CHAPTER 11

1. There are numerous textbooks at different levels that teach electromagnetism. For the introductory college level, see R. P. Feynman, *The Feynman Lectures on Physics*, vol. 2 (New York: Addison-Wesley, 1970); the standard graduate text is J. D. Jackson, *Classical Electrodynamics* (New York: John Wiley and Sons, 1999). The Uppsala University Department of Astronomy and Space Physics offers a free downloadable text in classical electrodynamics, as well as numerous links to related sites, at Bo Thidé, "Classical Electrodynamics," http://www.plasma.uu.se/CED/ (accessed June 11, 2004).

2. John P. Ralston, personal communication with CTH, May 1996; see also J. D. Jackson and L. B. Okun, "Historical Roots of Gauge Invariance," *Reviews of Modern Physics* 73 (2001): 663. Jackson and Okun write:

> Notable in this regard, but somewhat peripheral to our history of gauge invariance, was James MacCullagh's early development of a phenomenological theory of light as disturbances propagating in a novel form of the elastic ether, with the potential energy depending not on compression and distortion but only on local rotation of the medium in order to make the light vibrations purely transverse. . . . MacCullagh's equations correspond (when interpreted properly) to Maxwell's equations for free fields in anisotropic media. We thank John P. Ralston for making available his unpublished manuscript on MacCullagh's work.

Thus MacCullagh actually constructed a theory of light as a propagating wave disturbance in a material medium, an "ether," in 1839. This theory is equivalent to Maxwell's theory of some twenty-five years later, and it involves the concept of an unobservable gauge field; hence, MacCullagh seems to have discovered the symmetry principle of local gauge invariance. But the connection of the underlying physical picture here, involving the concept of twists, or local rotations, in a material medium, to electrodynamics is remote. This discovery has gone almost completely unnoticed. MacCullagh, whose relationship with the rest of the physics community was not a happy one and whose life ended tragically in suicide, may have been a man too far ahead of his time.

3. In particular, a *complex phase factor* is just an exponential, like $e^{i\theta}$, where θ is real, and this has a magnitude of unity, i.e., $1 = |e^{i\theta}|^2$. So multiplying the electron wave function by this factor means $\psi(\vec{x},t) \rightarrow e^{i\theta} \psi(\vec{x},t)$; this doesn't change the magnitude of the electron's wave function,

and it therefore shouldn't affect the measured probabilities. The key to *local* gauge invariance is that we will allow the angle appearing in the phase factor to be a *real function* of space and time, $\theta(\vec{x},t)$. This can change the apparent energy and momentum in the electron wave function. That is, given that our electron has a momentum $\vec{p}$ and energy E, it is easy to find a $\theta(\vec{x},t)$ such that, if we multiply the wave function by the phase factor $e^{i\theta(\vec{x},t)}$, we will then have any *new* momentum $\vec{p}'$ and energy E' that we want. For example, with the simple choice $\theta = ax - bt$, then we find $e^{i(kx-\omega t)} \rightarrow e^{i(ax-bt)} e^{i(kx-\omega t)} = e^{i(k'x-\omega't)}$, where now $k' = k + a$ and $\omega' = \omega + b$. We've evidently changed the wave number and frequency in an arbitrary way. Thus, multiplying by a complex phase that depends upon space and time has turned the old electron, with its momentum $\vec{p}$ and energy E, into a new one, with an arbitrarily different momentum $p' = \hbar k'$ and energy $E' = \hbar\omega'$. This apparently is not a symmetry of the original state, but rather, it has produced a new state, with different observable energy and momentum.

4. This new gauge interaction changes the total energy of the electron by providing an additional potential energy. Then the total energy of the electron is modified: $E = \hbar\omega + e\phi$. And, since special relativity tells us that under boosts the energy and momentum must mix with each other, like time and space, it follows that we will also have to introduce something new, which modifies the momentum in a similar way: $\vec{p} = \hbar\vec{k} + e\vec{A}$. These new objects, $(\phi, \vec{A})$, act like time and space $(t,\vec{x})$ under Lorentz transformations. We call ϕ the *scalar potential* and $\vec{A}$ the *vector potential*. The constant, e, is a factor called the *electric charge* and determines how strongly the electron will feel the presence of the new scalar and vector potential. Now, when we multiply our electron wave function by the phase factor, which shifts the frequency and wave vectors to new values, $\omega \rightarrow \omega'$ and $\vec{k} \rightarrow \vec{k}'$, we can simultaneously shift the values of the new scalar potential and the new vector potential. That is, the overall gauge transformation is as follows: (1) multiply the electron wave function by the phase: $e^{i(kx - \omega t)} \rightarrow e^{i\theta(\vec{x},t)} e^{i(kx - \omega t)} = e^{i(k'x - \omega't)}$, and (2) then we see that $\omega \rightarrow \omega' = \omega + b$ and $k \rightarrow k' = k + a$, but (3) we also shift the scalar potential: $\phi \rightarrow \phi - \hbar b/e$, and (4) we shift the vector potential: $\vec{A}_x \rightarrow \vec{A}_x - \hbar a/e$. Then we find that, under the combined transformations, the total energy is unchanged: $E = \hbar\omega + e\phi \rightarrow \hbar\omega' + e\phi - \hbar b \rightarrow \hbar\omega + e\phi = E$. In addition, the momentum is unchanged under the combined transformation as $\vec{p}_x = \hbar\vec{k} + e\vec{A}_x \rightarrow \hbar k' + e\vec{A}_x - \hbar a = \hbar k + e\vec{A}_x = \vec{p}_x$. These combined transformations are called a *local gauge transformation*.

CHAPTER 12

1. For lack of space, we regretfully won't discuss the periodic table in detail. Many versions can be found on the Internet, such as "A Periodic Table of the Elements at Los Alamos National Laboratory," http://pearl1.lanl.gov/periodic/default.htm (accessed June 14, 2004). Modern tables include nonnaturally occurring and recently synthesized elements, through atomic number 118.

2. One electron volt of energy corresponds to 1.60×10^{-19} joules, and this is equivalent, dividing by the speed of light squared, $m = E/c^2$, to a mass of 1.78×10^{-36} kilograms. If we multiply the proton mass of 0.938 GeV (billion electron volts) by 1.78×10^{-36} kilograms we get 1.67×10^{-27} kilograms, the proton mass as measured in kilograms.

3. See, for example, Murray Gell-Mann, *The Quark and the Jaguar* (New York: W. H. Freeman, 1994); this is not a biography but a very intriguing treatise on complexity, physics, and other issues.

4. Gell-Mann proposed the term *quarks*, borrowed from the passage in James Joyce's novel *Finnegan's Wake*: "three quarks for Muster Mark," thankfully breaking the tradition that everything requires a Greek alphabetical symbol for nomenclature in particle physics. The idea of quarks was also independently proposed by George Zweig, a colleague at Caltech of Gell-Mann, who happened to be on a visit to CERN. Zweig wrote down the idea in a famous unpublished CERN preprint. Zweig chose the name *aces*. He realized that certain dynamical properties of the many newly discovered particles could be explained on the basis of this next layer of matter, the quarks.

5. We actually have some understanding of the fact that the electric charges of the leptons and quarks of a given generation must add to zero. This comes from the requirement of a cancellation of the "Adler-Bardeen-Bell-Jackiw" anomalies, a quantum threat to the gauge symmetries in the weak interactions. This "anomaly cancellation" occurs most easily for the particular pattern of quarks and leptons that are seen in each generation. We also have beautiful and compelling "unified theories," such as the Georgi-Glashow $SU(5)$ theory, that "predict" this pattern. However, in any theory, we can't be absolutely sure that a particular lepton, e.g., the electron, necessarily goes with the up and down quarks, as opposed to the top and beauty quarks or some other scrambling and rearranging of things.

6. However, before we get too carried away as cost-cutting particle

efficiency experts, intending to trim nature's generations of particles, let us remark that the CP violation observed in nature *does require* all three generations, for technical reasons, and we also have seen that some kind of CP violation is necessary for matter to exist in the universe at all. In addition, all quarks and leptons were active in the early universe and played a role in the formation of the universe we ended up with. We would therefore be remiss to discard them.

7. The generational pattern actually involves the helicities, or more properly the *chiralities*, of the quarks and leptons. That is, only the "left-handed" particles couple to the weak force in any generation. One could have a continuation of this pattern, but it would require both "right-handed" and "left-handed" particles coupling to weak interactions.

8. Taken together, we refer to all the multitude of composite particles that we can build out of quarks (and antiquarks) as *hadrons*. Those hadrons composed of three quarks (or three antiquarks) are called *baryons*, whereas those composed of a quark-antiquark combination are called *mesons*. There are corresponding "excited states" of these quark-containing particles, called *resonances*, which behave like various quantum energy levels, such as the electron trapped in a potential well, exhibiting the "guitar string modes" of chapter 10. The baryons all have spins of 1/2, 3/2, 5/2, etc.; the mesons all have spins of 0, 1, 2, etc.

9. The number of gluons is $8 = 9 - 1$. The number of (color, anti-color) pairs that we can logically have is nine. One combination, $r\bar{r} + b\bar{b} + g\bar{g}$, is not an $SU(3)$ symmetry group element. That is, it doesn't rotate anything in color space and doesn't arise as a gluon. This leaves a total of eight physical gluons, whose effects have been dramatically observed in jets produced in high-energy collisions.

10. Some author has even referred to the Higgs boson as the "God Particle," because of the profound impact it has upon us and the whole universe.

APPENDIX

1. There are other branches of mathematics, such as *homotopy*, that are concerned with the paths we can take on different surfaces, or in different spaces, and how many times we wrap around obstructions of holes, etc. This is a branch of topology. For example, consider all possible closed curves that start and return to a point, P, on the surface of a sphere.

We consider that all curves that can be *continuously deformed* into one another (provided we don't break or cut the curve) are equivalent. Then we see that all curves that pass through P are equivalent on the sphere, because any curve can be deformed into any other. However, suppose we are on the surface of a doughnut. Then curves that wrap N times around the doughnut like the whitewall on a tire are *inequivalent* to curves that wrap M times, when $N \neq M$. Also, curves that wrap Q times around the doughnut like the "radial" of a tire are *inequivalent* to curves that wrap L times, when $L \neq Q$. Therefore, all curves on a doughnut are distinguished by a pair of *winding numbers*, (N, Q), that determine how many times they wrap in the whitewall or the radial direction. The fancy language is that the *homotopy group* of the doughnut is $Z \otimes Z$, the *cartesian product* of the set of all integers, Z, with itself. The homotopy group of a sphere is *trivial*, the set containing only the identity.

2. See D. Gorenstein, "The Enormous Theorem," *Scientific American*, December 1985, p. 104.

3. Sophus Lie had a hard time convincing his peers how significant his algebra of continuous groups was, and he ultimately went mad; see J. J. O'Connor and E. F. Robertson, "Marius Sophus Lie," http://www-gap.dcs.st-and.ac.uk/~history/Mathematicians/Lie.html (accessed June 18, 2004).

INDEX